I0488814

DIGIPERSON. Copyright © 2016 By Richard Ponschock and Gerard Becker. All
rights reserved.
Printed in the United States of America. No part of this book.
May be used or reproduced in any manner whatsoever without written
permission except in the case of brief quotations embodied in
critical articles and reviews.

ISBN-13: 978-1519493507 (CreateSpace-Assigned)
ISBN-10: 1519493509

The Sociological Impact of the Digital Persona: Mankind's Transformational Odyssey

Prologue

Walking down any busy 21st century street yields most pedestrians tethered to technological devices and sociologically interacting in ways that are much different than just a few years ago. Over the past several years, a series of research articles has resulted in this socio-technological odyssey herein. The attendant research was both phenomenological in nature while also longitudinal, spanning a period of five to ten years of work (Appendix A). The 21st century society is on a journey that could be equated to an amusement park roller coaster. Every turn and flip brings about new excitement, thrills, and sometimes fear. This book, DIGIPERSON: A Socio-Technological Odyssey will take us on a journey that will eventually arrive at the 21st century amusement park. The journey is not without some additional attractions of its own. These side trips will be explained chapter by chapter. First, let's look at this odyssey at a macro level, the big picture.

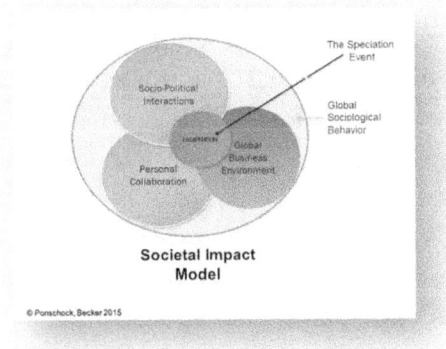

Figure 1: Macro Societal Impact Model

Woven into this excursion's overarching theme are three seminal social concepts impacted by the digital influence on 21st century society. These three impact points are our socio-political interactions, business relations on a global scale, and very specifically, our personal communications and interactions. These social impact points contribute along with the advancements in technology to the transformation of mankind into a DIGIPERSON. Technology has changed our behavior in every way imaginable; even the way that we travel along this transformational odyssey. This book is intended to be enlightening and educational. The end (or is there an end?) is eye opening.

The odyssey spans twelve distinct excursion points segmented into the three domains highlighted in the preceding illustration. A journey spanning man's debated beginning to an alarming current state of existence. It is difficult to illustrate a path that has traversed such a diverse terrain. This odyssey is undoubtedly taking us through some uncharted territory; but it is unmistakably heading toward a major event, which is termed a *speciation* event for this odyssey. The 21st century dweller is changing the way they communicate with each other. There is a growing avoidance of face-to-face communications in casual, business, and educational settings. There are technical drivers that are accelerants to the social behavioral shift away from personal communications. The cell phone, SKYPE™, and apps such as Facebook™ appear to be more commonly used than just meeting each other on the street corner or coffee shop. Even the coffee shops now have drive-thru capabilities, reducing *coffee shop* type discussions.

Transitioning from personal collaboration, business and economies have become intertwined into a global network of interacting forces. These forces flow into geopolitical issues where even wars and territorial battles have an impact globally and are executed via technological advances. Global business interactions are now conducted with minimal, if any, personal interface in many cases. Even the advent of potentially global currencies is evolving via these technological capabilities and is explored as instrumental components for the impending *speciation* event.

As the odyssey unfolds, the potential sociological reconstruction of this current day culture in future years will not be about digging for remains with a pick and shovel; rather, future archaeologists will attempt to reconstruct this current day culture via the unearthing of all the data that is left behind. A significant aspect of this odyssey results in a discussion related to what has been termed herein as *"Digital Identity Crumbs"*[©].

The following figure depicts a micro view of the impending *speciation* event with the attendant forces that lead towards that event. Consider some of these drivers as....

- Artificial intelligence;
- Bio-technology;
- Cloud technology;

- Information Computer Technology Management (ICTM); and
- Other technologies

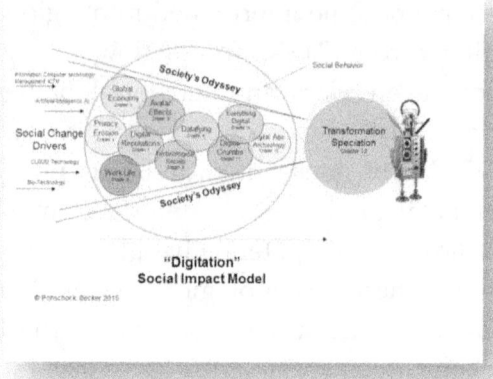

Figure 2: Micro Speciation Social Impact Model

So, *"buckle your seatbelt"*, climb aboard the roller coaster for this transformational odyssey, and be prepared to experience a genuine paradigm shift in personal, organizational, and inter-cultural relationships based on a new model of individual termed herein as the DIGIPERSON. It is now time to embark on that journey......

Table of Contents

I truly believe this is a century unlike any other because this is a century in which everything physical and analog will become digital, mobile, virtual, personal.
Carly Fiorina,
Former Chairman and CEO, Hewlett-Packard

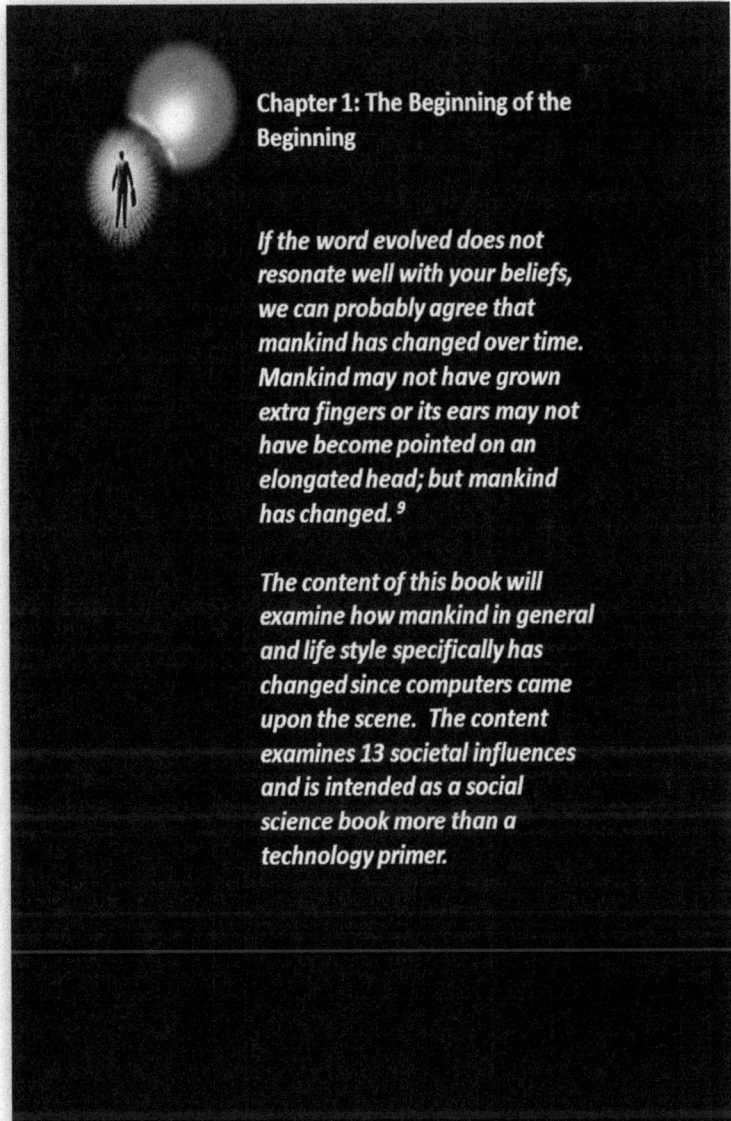

Chapter 1: The Beginning of the Beginning

If the word evolved does not resonate well with your beliefs, we can probably agree that mankind has changed over time. Mankind may not have grown extra fingers or its ears may not have become pointed on an elongated head; but mankind has changed. [9]

The content of this book will examine how mankind in general and life style specifically has changed since computers came upon the scene. The content examines 13 societal influences and is intended as a social science book more than a technology primer.

Background

The "Digitization of Society: The Transformation of Humankind into a Digital Person" is a synthesis of Sociological, Anthropological, Psychological, and Biological theory. Without much argument, each of these disciplines provides substantive contributions to the study of mankind and the societal environment in which they thrive. Therefore, in order to understand the digitization of mankind and the development of the DIGIPERSON, one must have a fundamental understanding of the beginning, the history, and the social impacts that laid the path to where mankind is today. This book also provides a potential glimpse into the futuristic state of mankind which might occur in a very short period of time. The content of this work is a compendium of a longitudinal study on the impact that *digitization* has had, and is having, on human society.

The Study of Life

The origin of life has been, and continues to be studied from multiple perspectives and disciplines. Physicists currently carbon date artifacts of rocks and speculate on the 'Big Bang' theory. Biologists continually attempt to determine which bacteria may reproduce. This book takes the reader on a journey along the digital era and continuum, while illustrating the impact that *digitization* and computerization of modern man's surroundings has had on society, and civilization at-large.

Modern Man

For the purpose of this book, we fast-forward to MODERN man and begin the analysis at or around the onset of the computer era. The content examines 13 societal influences which are illustrated in the social impact model.

Figure 1: The social Impact Model

"Digitation"
Social Impact Model

© Pomschock, Becker 2015

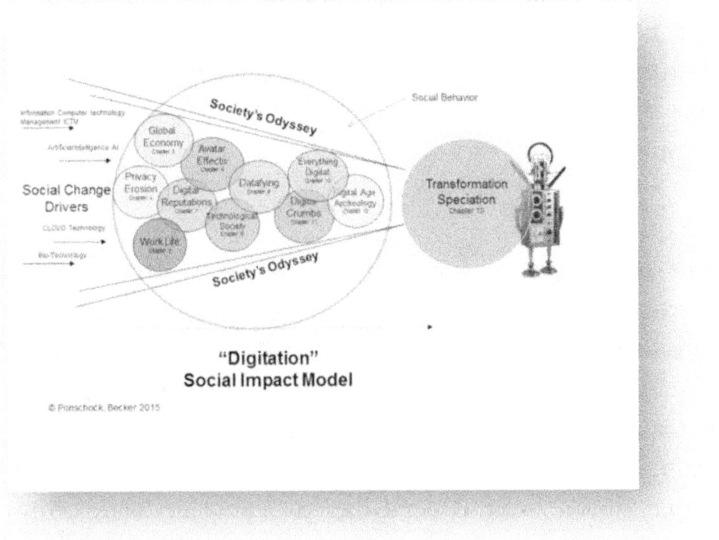

Chapter 2

Establishes a continuum on which the societal impact model is built. Information technology has been in a transitional (and somewhat transformational) mode since its arrival on the business landscape. Contributing to a rapidly changing position that computer technology plays in the 21st century, the overall population and fast paced dynamics surround computerization itself. The content is not a study in technology; rather it is an examination of the digital impact on the societal life style and existence.

Chapter 3

As digital transformation evolved, the world economy became intertwined. Country by country economies became interdependent with increasing impact on each other. The stock market in Asia affects the opening bell in the United States. The knitting of individual economies has created a virtual society where borders, currencies, cultures, and dependencies are now blurred. The nature of this distortion can be viewed as virtualization; a blending of multiple realities into one.

Chapter 4

As traditional boundaries collapse and digitization increases, privacy and personal rights are further impacted. Although this book primarily focuses on the United States and its legal system, the fundamentals can be viewed through the lens of the European Union as well. This chapter examines how the digital environment impacts humankind forming a "new normal" from a privacy perspective.

Chapter 5

Connected to the "New Normal" on privacy is the work model that many 21st century workers are experiencing. Mobility, and the lack of separation between work, family, and play has created what the authors refer to as the work from anywhere model. Staying continually connected has impacted our very existence. Some cities have special "cell phone" lanes on sidewalks to mitigate injuries from individuals colliding due to lack of concentration. Laws are being created to prohibit texting and driving. Working from anywhere is another "new normal".

Chapter 6

Individual identity is under siege. To combat the erosion of personal privacy and personal rights, many of the users of social media and 21st century's social chatter now create multiple "virtual" images of themselves. As society in general is transforming digitally so is the persona of the individuals making up that society. The human inhabitants of the 21st century are taking on multiple identities. This chapter examines the avatar effect, an alternate reality, on the digital dwellers. The use of multiple "We" can be labeled as the "Avatar effect".

Chapter 7

As discussed in the previous chapter, the "Avatar affect" can change our persona; the digital/cyber world can also jeopardize our reputation. Identities are stolen every 2 seconds and reputations get tarnished without the victims' knowledge. Business and individuals continue to stay on guard to ensure that information shared in the cyber world does not tarnish our character in the physical world.

Chapter 8

The physical world and the digital landscape are becoming Social networks. The transformational processes are integrating employees, customers, and suppliers that were not even contemplated five to ten years ago. As Prahalad & Krishnan note "There is a fundamental transformation of business underway. Forged by digitization, ubiquitous connectivity, and globalization, this will radically alter the very nature of the firm and how it creates value..."

Chapter 9

Acknowledging the redundancy in the use of the following Prahalad & Krishnan quote, "There is a fundamental transformation of business underway. Forged by digitization, ubiquitous connectivity, and globalization, this will radically alter the very nature of the firm and how it creates value..." This chapter dives deeper into DATIFICATION by exploring the next phenomenon, *The Internet of Things* (IoT). The previous discussions dealt with man and machine and their integration with society; IoT provides and analysis and discussion on machine to machine monitoring and interaction. Common *machines* like refrigerators and microwaves are now making decision; automobiles park themselves while also braking if encountering objects in their paths. This is only the beginning of further DATIFICATION.

Chapter 10

Subsequent to the exploration of DATAFICATION on general terms, this chapter examines society through the *datafied* lens. As a society, we're producing and capturing more data each day than was seen by everyone since the beginning of the earth.[17]

Chapter 11

While the previous chapters illustrated how we interact on the digital landscape, Chapter 11 examines the digital footprints and digital DNA that leave everlasting imprints. The digital remnants of our daily actions have been termed "Digital Identity Crumbs ©".

Chapter 12

Digital footprints, new social media, electronic waste requires new skills for archeologist in the future. Anthropologists analyzing the 21st century will use digital mining techniques, computers, software programs and electronic instrumentation, and the potential need to reconstruct the computer operating environments, data formats, etc. required to meticulously recreate, read and interpret the digital information.

Chapter 13

During the last century we have observed Dolly the cloned sheep. Dolly's existence was announced to the public on 22 February 1997.[6] Dolly was the first clone produced from a cell taken from an adult mammal. The production of Dolly showed that genes in the nucleus of such a mature differentiated somatic cell are still capable of reverting to an embryonic state, creating a cell that can then develop into any part of an animal.[11]

Organovo® has already started developing a 3D-printed liver model for testing the safety and efficacy of drugs. The startup company is also creating cancerous versions of living tissue models for testing cancer drugs. Eventually bio-printing will revolutionize the delivery of tissue. This technology can develop into "tissue on demand" within the next 10 or 15 years.[8]

Society and mankind is now facing a new evolution; a crossroad that was never imagined outside of science fiction books and comics. Bioscience and technology has advanced to a level that makes it possible to modify, reconstruct, and even create living organisms. An athlete with artificial limbs ran track in the Olympics of 2012. We are a species that may be re-creating or re-building itself.

We are in the midst of an era that illustrates a speciation event may be imminent. The DIGIPERSON is on the horizon (or, is it already here?).

In summary, this journey book examines transformation, it's very basic definition, and how it differs in a virtual/cybernetic world. The book illustrates how mankind and its existent society is physically dependent on digital surroundings. The ability to artificially extend life can very well be the speciation event equivalent to the formation of the first bacteria that began to reproduce. So let's dive into transformation.

> *It is manifest that sociology must depend upon biology, since biology is the general science of life, and human society is but part of the world of life in general; it is manifest also that sociology must depend upon psychology to explain the interactions between individuals because these interactions are for the most part interactions between their minds. Thus on the one hand, all social phenomena are vital phenomena, and on the other hand, nearly all social phenomena are mental phenomenon.[15]*

Chapter Wrap-up

This chapter illustrated that regardless of personal beliefs in the origin of man, it can be agreed that the universe, the galaxy, the earth, and man has and is in the state of transition – a transformational change. Warmoththeorized "human intelligence evolved during the last ice age. The culminating phase of human biological evolution was intimately intertwined with the development of language and other forms of culture".[14] Biologists studying the theory of evolution hold the position that life hasn't always been as we see it today nor will it remain unchanged. This book examines how the "digitization" process is an accelerant to both further evolution and subsequent transformation. "Psychologists and sociologists use social context to study social changes over time and is grounded on the interplay between social forces that affect individual behavior that change society."[16]

End Notes

1. Cawdron, P_., (2012). Freeze-dried bugs. Retrieved May 29, 2014 from http://thinkingscifi.wordpress.com/2012/09/23/freeze-dried-bugs/
2. Culture, and Cloning". *Ethnos: Journal of Anthropology.*
3. Darwin, C., (1859). On the Origin of Species by Means of Natural Selection, or the Preservation of Favoured Races in the Struggle for Life,"
4. Darwin's Theory, (2014). Darwin's Theory Of Evolution - A Theory in Crisis Retrieved September 29, 2014 from http://www.darwins-theory-of-evolution.com/
5. Dobzhansky, T. 1937. *Genetics and the Origin of Species.* reissued 1982, with an introduction by S. J. Gould. ed. New York: Columbia University Press. 1956. *The Biological Basis of Human Freedom.* New York: Columbia University Press.
6. "Dolly the sheep clone dies young", BBC News, Friday, 14 February 2003 http://news.bbc.co.uk/2/hi/science/nature/2764039.stm
7. Farrell, J (August 27, 2010). "Catholics and the Evolving Cosmos". *The Wall Street Journal.* Retrieved September 11, 2012.
8. Hsu, J., (2013, September24). 3D Printing Aims to Deliver Organs on Demand. Retrieved October 9, 2014 from http://www.livescience.com/39885-3d-printing-to-deliver-organs.html
9. Meyer, C., Minnich, S., Moneymaker, J., Nelson, P., & Seelke, R., (2007). *Explore Evolution: The*

Arguments for and Against Neo-Darwinism, Hill House Publishers Pty. Ltd., Melbourne and London

10. Denton, M., "Evolution: A Theory in Crisis," 1986, p. 250.

11. Niemann H, Tian XC, King WA, Lee RS (February 2008). "Epigenetic reprogramming in embryonic and foetal development upon somatic cell nuclear transfer cloning". *Reproduction* **135** (2): 151–63. doi:10.1530/REP-07-0397. PMID 18239046.

12. Timeline, (2014) Timeline of evolutionary history of life. Retrieved May 29, 2014 from Retrieved http://en.wikipedia.org/wiki/Timeline_of_evolution ary_history_of_life

13. Vuletic, M., (1997). Charles Darwin, "On the Origin of Species by Means of Natural Selection, or the Preservation of Favoured Races in the Struggle for Life," 1859, p. 155. Review of Michael Denton's *Evolution: A Theory in Crisis* Retrieved October 11, 2014 from http://www.talkorigins.org/faqs/denton.html

14. Warmoth (2000) theorized "human intelligence evolved during the last ice age.

15. Ellwood, C. A., (1910). *Sociology and Modern Social Problems.* New York: American Book Co.

16. Briggs, J. (2012). Social Context Theory http://www.ehow.com/about_5414476_social-context-theory.html retrieved on 05/05/2013>

17. Conner, M. (2013, July 18). *Data on Big Data.* Retrieved from Marcia Cornner: http://marciaconner.com/blog/data-on-big-data/

Chapter 2: Technological Dynasties along the Odyssey

Change itself does not signify transformation. This chapter will examine ways the CLOUD has institutionalized an evolutionary change in the way societal communication takes place. The dialog will illustrate how user acceptance is being driven by the diffusion of technology. While the diffusion process of innovation [35] and the dynamics of the CLOUD may have influenced the growth of social media in volume and popularity; social networking has become a powerful force throughout the world. We've recently seen the power of Twitter and other microblogging tools in Egypt and Libya. The entire world uses these tools everywhere ... [43]

Background

Gharajedaghi submits the game is changing. The game Gharajedaghi focuses on is business in general and business architecture specifically. CLOUD computing may be the ambiguous driver of change.[14] "CLOUD" computing is the Internet-based, just-in-time delivery of data, movies, computer applications, storage, and computing power as services, done in a way that completely shields consumers from underlying technical details."[28] The commentary will explore the probable necessity for empirical research to clarify if the CLOUD architecture is the ends or the means on the ICT transformational continuum. Utility computing, the long-held dream in computer architecture, has the potential to transform a large part of the IT industry, making software even more attractive as a service and shaping the way IT hardware is designed and purchased. Developers with innovative ideas for new Internet services no longer require the large capital outlays in hardware to deploy their service or the human expense to operate it.[1] Are the changes that CLOUD technology is bringing, or will bring to the business landscape transformational or simply an evolutionary step to a broader progression? ICT has influenced the business landscape unlike any other innovation. Additionally, ICT riddled organizations with disruptive introductions from its inception. Gharajedaghi, advances that "ends and means are interchangeable concepts; an end is a means for further ends."[14] We contend that this cyclical principle lives on the Information Communication Technology (ICT) continuum. Figure 4, illuminates the evolutionary continuum of the business technology epoch, although unbroken, it has five distinct domains (enthralled, diffusion, mobility, and commoditization, ubiquitous).

These tremors resulted in both the end of the current dynasty and the beginning of the next. The question that arises: are these tremors a paradigm shift at the core of ICT or a transformation on how information will be governed and delivered? Kuhn authored the theory that a paradigm gains status because it is more successful than its predecessor. It is a solution to a problem that practitioners in a specific discipline view as acute.[23] The end of each dynasty is the beginning of the next advancement. The intent of this examination is exploratory. Will the CLOUD be more than just the beginning of a new ICT domain or will it be the accelerant for a technology transformation?

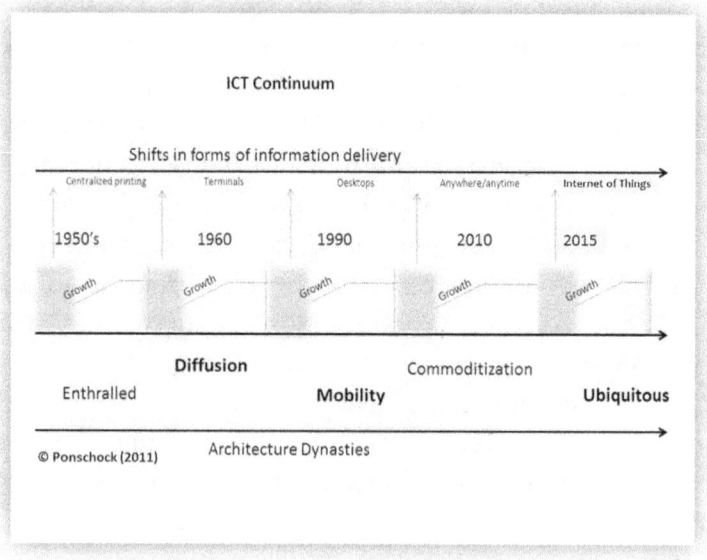

Figure 4: ICT Continuum

Historical Foundation

In 1936 Alan Turing, a 24-year old Cambridge University mathematician, invented the idea of what progressed to today's digital computer.[6] The actual beginning of the computer technology era for business applications can be argued. This chapter will consider the late 1950s as the beginning of the computer era for business. This is the period when data processing "tabulating" machines were transitioning the business landscape into the computer age. Information Technology has been in a continual state of evolution since the beginning of the computer age.[49] The acronyms attached to computer departments have evolved over time. The names ranged from Data Processing (DP) to the current tag ICT, an acronym for Information Communication Technology. Between these two designations were MIS, CIS, IS, and IT. (Ward, Peppard, 2009) Gordon E. Moore the co-founder of Intel set the advancement metric with his statement *"The numbers of transistors incorporated in a chip will approximately double every 24 months."* This simple phrase is now known as MOORES LAW.[29] Intel's silicon wafer set in motion a progression along the current continuum of technology evolution.

Dynasties Defined

Enthralled Dynasty

Overlapping the use of tabulating machines, the general purpose business computer era started with IBM's commercial offering of the IBM 1620 in 1959. Although a business machine, the IBM 1620 was still marketed as an inexpensive "scientific computer".[18] The System 360 was the first mainframe family of computers entering the business landscape and sold between 1964 and 1978. The IBM 360 series was designed to cover the complete range of applications, from small to large, both commercial and scientific.[19]

During the birthing years of business computing, computers were put into glass walled rooms so the status and stature of the company could be touted. Businesses were enthralled with computers both from their capabilities and potential. It was equally prestigious to utilize computers or be an employee with computer skills. From an organizational perspective, the computer department was both a productive and a disruptive force in the organization. The computer operators, programmers, and analysts were "behind the looking glass". Computers were typically in environmentally special rooms with glass walls.

Image 1:

Computer technicians needed special training and were almost set on a pedestal from an organizational perspective. They were generously paid and it was commonly said that they spoke a different "language".[11]

Diffusion Dynasty

In computer time the "enthralled" period was short lived and the large mainframe computers began sharing the landscape with the mini-computer, personal computer, and laptop. We refer to this as the "diffusion" dynasty. [35] For the first time, computer technology was moving into functional departments. Although the major technology tasks of computer operations and application programming were still in the hands of the computer specialists, transaction entry and simple inquiries were being migrated to functional personnel. Computer terminals were showing up outside of the computer rooms and on desks of accounting staff, and even being utilized on the manufacturing shop floor.

During the 1960s major shifts in computer architecture and use were occurring. The transition from room sized machines was soon replaced by minicomputers and then inexpensive personal computers (PC). The term minicomputer evolved in the 1960s to describe the "small" computers that became possible with the use of integrated circuits and random access core memory. The minicomputer usually took up one or a few large refrigerator size cabinets compared to the room size mainframes. The first successful minicomputer was Digital Equipment Corporation's PDP-8.[45]

Machine size alone did not make this era significant. With the minicomputer came the video display terminal (VDT).

Image 2:

The display terminal, though still tethered by cable, emancipated the business user from the data entry function of the computer department. The business user was finally permitted to enter, edit, and view their data. In addition to delivering information to a video screen, central reporting was also being diffused. Printing began occurring on demand and in the department that needed the output. Actual "hands on" computer equipment began to be diffused,"spread out," and "extended". The diffusion process of innovation and the dynamics that influence user acceptance of new information technologies is of interest both to researchers in a variety of fields as well as procurement of technology. Rogers punctuates the diffusion of innovations as the practice by which innovation is communicated over time among the members of a social system.[35, 40]

Mobility Dynasty

As the trend of diffusion progressed, the mobility dynasty became predominant. Although the personal computer was an important element of this time period, the cellular telephone, and other wireless devices gave all levels of society the ability to communicate and work across vast distances. Social media has grown in volume and popularity. The concept of virtual villages is being adopted by large corporations seeking community-building across networks. Like the popular Facebook, the corporate social networks allow employees to upload photos of themselves—not just corporate ID mug-shots, but personal information. A major bank envisioned a sophisticated knowledge-management platform integrated with multiple applications that facilitate access to information that is simple and intuitive. The network allowed easy access to other employees and departments showing their availability at any given moment. Through the network, members were able to search for best practices on marketing or sales topics, and conduct dialogue through blogs and wikis written by informed employees. Currently companies of all sizes public, private, and governmental have one or several of social media icons Connect on their web page. Shared services are being utilized to accomplish the tasks major corporations once tried to launch in-house.

Facebook has since documented 175 million active users [25] and is being used by many companies. COMCAST uses Twitter allowing the cable company's customers to ask service questions.[25] Conceptually, it will provide clearer understandings of who knows what and who knows whom, within a company and among its business partners. Social

31

networks have the capability to operate as central nervous systems[27] for organizations.

Commoditization Dynasty

The evolution of computers from programming systems to service systems is similar to the earlier evolution from the bare machines to those with the support of sophisticated operating systems, file systems, libraries, and other resource management facilities. The service environment provides the supplier the facilities needed to enable the presentation of services to its customers without requiring knowledge of the complexities of the computer system. In other words, a whole range of IT functions -- from e-mail to supplier relationship management -- can and should be thought of as commodity services. The next logical progression is to pick those services from a menu provided by a professional services firm.[21]

Contemporary authors and futurists are declaring the end of IT as we know it. Nicolas Carr in the Spring 2005 issue of MIT Sloan Management Review compares Information technology to electric power.[5] "...as a business resource, information technology today looks a lot like electric power did at the start of the last century [when manufacturers built and maintained their own generators]. He continues that "When overcapacity is combined with redundant functionality, the conditions are ripe for a shift to centralized supply".[5] As an example, according to Lucas Mearian in an article for CIO Magazine; IT shops still waste as much as 60% of their storage capacity.

There is no doubt that certain sectors of IT have been commoditized and the possibility that an internal IT organization can operate them better than a specialized provider is vanishingly small. An outsourced approach to these sectors is obvious. Tying up capital, operational budget, or precious headcount in supporting a commodity application is clearly wrong.[16]

Ubiquitous Dynasty

This is the era of unseen computers and digitization. Technology is hidden in the tools used in the delivery of everyday tasks and routines. These tasks are as simple as a morning cup of coffee that is ready because of a programmable coffee maker that starts the brew cycle ½ hour before wakeup time. Although you may not have made a conscious observation, the bathroom temperature was increased for the shower that you were about to hop into. This simple environmental change is the result of a programmable thermostat.

Shortly after a shower ended the sprinklers begin watering the lawn. No thought is given to its start time, but it occurred after the shower NOT during it. The timing prevented getting scalded because of the decrease in cold water pressure. A cell phone or automobile has more computing power than the space vehicle of Apollo 13. The car has a security system, an emission control system, satellite radio, 4G Internet connectivity, remote start capabilities and many more options. The cell phone has replaced the personal digital assistant (PDA), the camera, the pager, computer games, and even have the ability to watch your favorite football game if it is not being televised on your local channel. There are thousands of applications "APPs" downloadable in seconds. The cell phone can tell the temperature, traffic congestion, and alternate driving directions.

While stopping for gas, a credit card is swiped for payment but not before a bar code is scanned for the bottle of water purchased. After a day of work, the smell of a roast beef greets you the result of a programmable slow cooker or microwave. While away at the office, the pool pump came on to perform a daily cleaning. Figure 5 lists many of the ubiquitous computers affecting daily life.

Security	Daily Tasks	Kitchen	Auto	Health	Cell Phone
Alarm system	Irrigation	Microwave	Emission	Pacemaker	Voice
Smoke detectors	Pool Cleaning	Refrigerator	Tire Pressure	Digital scale	Text
Heath alert pendants	Debit cards	Coffee maker	Satellite Radio	Diabetes monitor	Camera
Remote door locks	ATMs	Stove	4G service	Activity monitor	GPS
	Check deposits by cell phone		Intrusion detection	Hearing aide	Television
	Thermostat		Remote start		Numerous Apps
Etc.	Etc.	Etc.	Etc.	Etc.	Etc.

Figure 5: Ubiquitous Computers

The ubiquitous nature of technology has been given a name and an acronym. The Internet of Things (IoT) a term that refers to the concept that more than just people can connect to the Internet. In the era of today's technology many devices are in a continual state of connection. Technological advancements are leading to the creation of new products and services that rely upon connectivity to operate. This continual state of connectivity means that data is being sent to and from these devices at all time. A recent study conducted by eMarketer found that "Worldwide Smartphone Usage to Grow 25% in 2014," and "Nine countries to surpass 50% smartphone penetration this year."[10] The smart phone saturation represents a very obvious category of the Internet of Things. From a marketing and business standpoint, reaching these smart phone users is essential. However, the Internet of Things is not something that is only business related. It is something that is revolutionizing everything around us from the farming industry to medical care, and automotive vehicles to military tactics. With this diversity it is important to understand the advancements in connectivity, the driving factors of the Internet of Things, the future of devices, and potential negatives. Understanding where this data is coming from will help us learn what to do with it all. This book has a deeper discussion on IoT its implications and impact on the digitization of mankind in a later chapter.

CLOUD Architecture

The term "cloud computing" has gained a contemporary meaning in the information technology world in the past 18 months as a way to describe the ongoing evolution in how people access and manage digital information.[17] Look for a moment at the cellular phone; this is an example of a very complex device that started as a dedicated limited set of services (i.e. making an audio connection to another person at a distant location). When placing a telephone call, a subscriber is using a very complicated communications network; the caller simply dials the number of the phone he wishes to be connected to.[12] Since Frankston made the observations in 1974 the "cell" phone can connect to shared-services that range from dedicated mobility applications to social networking interactions. The information they seek on mobile devices is practical and real time: 42% of mobile device owners (i.e. cellular phones, tablets) obtain weather updates and 37% get material about restaurants or other local businesses. Fewer get local traffic, news alerts or other local topics.[46]

Some argue that the CLOUD architecture is "back to the future", that CLOUD computing = mainframe time share? It can be agreed that time-sharing is the joint use of a computing resource among many organizations through the architecture of multitasking. Cloud computing refers to the applications delivered as services over the Internet. The mainframe hardware and operating software and network components in the data centers provide those services.[1] Cloud computing has been gaining increasing attention from businesses of all sizes as a way to obtain secure access to advanced technology that is not necessarily owned or hosted by the user.[20] CLOUD technology is profoundly changing the ICT architecture. As we have already observed, ICT architecture has changed several times throughout its existence. The CLOUD is only one variant in the changing ICT sphere. The CLOUD has been enabled and accelerated by the ubiquitous connectivity of the Internet. The computer architecture may be the accelerant but is it the fuel? Lucas in his book Inside the Future: Surviving the technology revolution states:

> *The technology field is associated with much exaggeration, from vendors' promises to the inflated expectations of the dot-com Internet stock market bubble. Is this notion of an IT revolution and transformation more of the same? How does one determine the difference between a true transformation and incremental change?*[24]

Gartner, a leading research firm furthers this assessment. Gartner placed CLOUD computing at the top of their 2010 "Hype Cycle". As shown in Figure 6, many of the variants of CLOUD computing are achievable in the next 2 to 5 years.

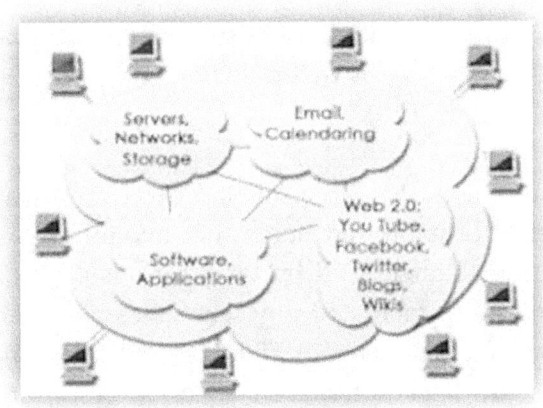

Figure 6: CLOUD Architecture

We argue that CLOUD computing may not be the *ends* of the current ICT architecture but may be the *means* for the industrialization or commoditization of the current information delivery platform.

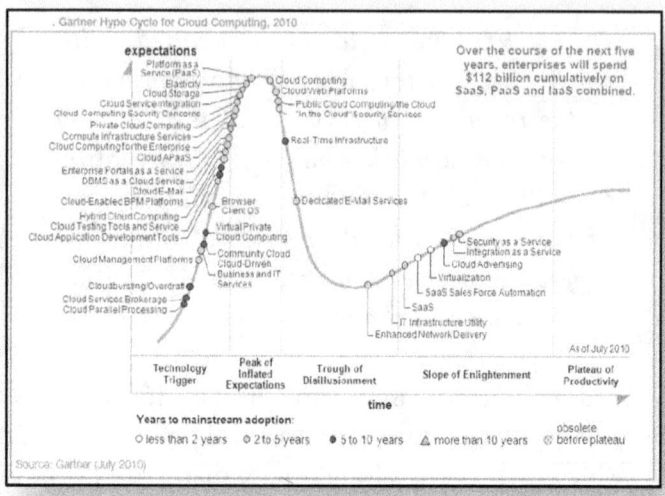

Figure 7: The Gartner Hype Chart[13]

The Gartner Hype chart would support the judgment that CLOUD technology is in the incremental - evolutionary epoch (Gartner, 2010).[13]

The theory of evolution holds that evolutionary change tends to be characterized by long periods of stability, or equilibrium, punctuated by episodes of very fast development. Referring to Figure 4, the ICT continuum models the evolutionary theory or punctuated equilibrium. In an empirical study of punctuated equilibrium, Romancelli and Tushman defined revolutionary transformation as one that occurred when they found changes in three major organizational conditions – strategy, structure, and power – within a two year period.[36]   Eldredge and Gould assert that the majority of morphological evolution occurs at or around the time of speciation events and that between speciation events there are long periods of evolutionary stasis during which no significant evolutionary change occurs.[8] Although the mobility dynasty continues to grow in usage and functionality, the technology is in a stable phase.   From data center consolidation to storage virtualization, CLOUD technology has a strategic place in ICT and corporate planning. The CLOUD is potentially a force of technological revolution and social change.

Berger and Luckman advanced the theory of social construction when they noted that everyday life is "not only taken for granted as reality by the ordinary members of society in the subjectivity meaningful conduct of their lives, it is a world that originates in their thoughts and actions and is maintained as real".[4] A major focus of social constructionism is to uncover individual behaviors and group participation in the construction of their perceived social realism. It involves looking at the ways social phenomena are created, institutionalized, known, and made into common practice by individuals. In the late 20th to the early 21st century, individuals have become fast paced and time-constrained, working longer and harder than they did 30 years ago.[42] Technology has helped fill some the time gap. A shift is occurring in the way members of our technologically-oriented society interact,[30] changing from one of personal relationships to a dialogue between individuals. Technology has driven the "new media" over the past 10 years. A distinguishing attribute of a true technological reconstruction is that many innovations occur at the same time. [22] Kuhn explained this shift as a paradigm change.[23] Technology is a complex system in which the actors construct artifacts in a context shaped both by their interests and by the underlying physical nature of their artifacts.[15]

In an earlier study, Ponschock discussed his findings on social shifts of Internet networking. His research indicated that the Internet creates an atmosphere where every communication is treated as if it were constructed in a small hometown – where everyone knows everyone else's business. As he noted, communities like MYSPACE are reflections of this migration to a virtual relationship.[32]

We propose that one of the speciation events in the 21st century is social networking in which the CLOUD is both the ends and the means. Social networking is a powerful force throughout the world. We've recently seen the power of Twitter and other microblogging tools in Egypt and Libya. The entire world uses these tools everywhere.[43] Social networking has fueled the need for the "plug-n-play" ease to connect to the CLOUD. Beniger a seminal thinker on society and evolutionary control posits:

> *As in earlier revolutions in matter and energy technologies, the nineteenth-century revolution in information technology was predicated on, if not directly caused by, social changes associated with earlier innovations. (p.9) "Because technology defines the limits on what a society can do, technological innovation might be expected to be a major impetus to social change.[3]*

40

Social networking enabled by Internet and CLOUD infrastructures is performing crucial roles throughout society. Just a few of the very positive purposes social networks can be used for include:

- Researchers can collaborate between any location in the world;
- Friends / families are no longer bound by distance;
- Academics can use Social networks to enhance courses and knowledge transfer;
- Businesses can use social networks internally and along the supply chain.

While the electronic form of social interaction has its limitations - the convenience and ease-of-use of social networks through blogs, forums, and e-mail-lists can lead to a society that is actually much more connected than ever before. Social media has now allowed this sphere of friends and colleagues to become global. These social networks will only serve to enhance relationships, and to benefit society as a whole. In doing so, these users are making use of "cloud computing," an emerging architecture by which data and applications reside in cyberspace, allowing users to access them through any web-connected device.[17] The age of the global community is upon us and enabled through advancements with technologies such as ICT. Consider as evidence the global classroom consisting of individuals around the world cooperating towards their desired objectives in real time!

Chapter Wrap-up

Change by itself does not signify transformation. A deluge of articles about CLOUD computing populate the

41

commercial publications and even television advertisements. The writers and the publishers focus on the value, dangers, strengths, and weaknesses of virtualization and the CLOUD. As far back as 1981, it has been theorized that the impact of business technology on the organization needed to be understood. [34] According to a 2003 survey, about 60% of the average U.S. companys' IT staffing budget goes to routine support and maintenance functions.

> *"When overcapacity is combined with redundant functionality, the conditions are ripe for a shift to centralized supply. Yet companies continue to invest large sums in maintaining and even expanding their private, subscale data centers. Why? For the same reason that manufacturers continued to install private electric generators during the early decades of the 20th century: because of the lack of a viable, large-scale utility model. But such a model is now emerging..."[5]*

Information technology has been in a transitional mode since its arrival on the business landscape. Contributing to this unsettled position that IT plays in the organization are the fast paced dynamics that surround computer technology itself. Intel, a leader in circuit chip manufacturing has delivered the challenge of Moore's Law: to double the transistor density, while increasing functionality and performance and decreasing costs. Social networks integrating employees, customers, and suppliers were not even contemplated five to ten years ago. Yet, if today's leaders and managers overlook these virtual connections, they will be doing a disservice to their organizations.

As Prahalad & Krishnan note, "There is a fundamental transformation of business underway. Forged by digitization, ubiquitous connectivity, and globalization, this will radically alter the very nature of the firm and how it creates value".[33] Supply chain partnerships between customers and suppliers are morphing into global, complex, interdependent exchange points, forcing organizations to extend planning beyond the four walls of their enterprise.[44] Many firms find themselves unprepared to accept the challenges posed by this new reality as managers and leaders face deeply engrained organizational legacies steeped in both social and technological agendas.[33] Management in the 21st Century plays a pivotal role in transforming mindsets, skills, behaviors, and decision structures of organizational leadership. This transformation to a virtual social architecture will only be successful through a period of socialization and acculturation where the organization and its component parts work together organically to effectuate transformative change. This exploration is surfacing a need for empirical research on a minimum the following themes.

1. Will Facebook, MySpace, and Twitter or the like become tomorrow's e-mail?
2. Will Google apps or the like replace the Microsoft office suite or with the Microsoft office suite become CLOUD based?
3. Will hosting of licensed software applications replace in-house implementations?
4. Will the computer careers and organizations of today be subjected to drastic transformation?

5. Will academia react quickly to provide the next generation of computer technician the proper foundation?
6. Is the CLOUD a transformational event in ICT architecture and,
7. Is Social Networking a Transformational event driven by the CLOUD?
8. Is society in the midst of a speciation event?

Technology experts and stakeholders say they expect they will 'live mostly in the cloud' in 2020 and not on the desktop, working mostly through cyberspace-based applications accessed through networked devices. This will substantially advance mobile connectivity through smartphones and other Internet appliances. Many say there will be a cloud-desktop hybrid.[47]

Cloud computing has been gaining increasing attention from businesses of all sizes as a way to obtain secure access to advanced technology that is not necessarily owned or hosted by the user.[20] Some 69% of online Americans use webmail services, store data online, or use software programs such as word processing applications whose functionality is located on the web. In doing so, these users are making use of "cloud computing," an emerging architecture by which data and applications reside in cyberspace, allowing users to access them through any web-connected device. [17]

Since 2005 CLOUD computing has gained momentum both in adaptation and industry press. Cloud computing promotes a new dialog between business and IT decision makers to define business service requirements first and the decide how to balance the use of shared, internal virtualized IT resources and external public services most cost effectively while maintaining required levels of cost performance, security, and business resilience. The focus on business service priorities results in better IT resource utilization. Cloud computing drives breakthrough improvements in IT Service Delivery, Speed, and costs.[48] Microsoft's Ray Ozzie has said, cloud computing would enable "a personal mesh of devices – a means by which all of your devices are brought together, managed through the web as a seamless whole."[31]

End Notes

1. Ambrust, M., Fox, A., Griffith, A. D., Katz, R., Konwinski, A., Lee, G., Patterson, D., Rabkin, A., Stoica, I., & Zaharia, M. (2010, April). A view of cloud computing. *Communication of the ACM.* 53(8), 50

2. Bell, G., Cady, R., McFarland, B., Delagi, J., O'Laughlin, Noonan, R., & Wulf, W. (1970). A new architecture for mini-computers- DEC PDP-11. *AFIPS Conference Proceedings,* Vol. 36 pp. 657-675.

3. Beniger, J. (1986) Havard University Press Cambridge MA The Control Revolution: Technological and Economic Origins of the Information Society

4. Berger, P.L., & Luckman T. (1967), The social construction of reality: A treatise in the sociology of knowledge, Garden City, NY: Doubleday.

5. Carr, N. G. (2005, Spring). The end of corporate computing. *MIT Sloan Management Review.* 46(3).

6. Carr, N. G. (2008). IT in 2018: From Turing's machine to the computing cloud. IT management eBook, Jupitermedia Corp

7. Cone, E. (2007). Social networks at work promise bottom-line results. Retrieved April, 4, 2011, from http://www.cioinsight.com/article2/0,1540,2192575,00.asp

8. Eldredge, N. & Gould, S. (1972) Punctuated Equilibria: An Alternative to Phyletic Gradualism, in T.J. Schopf (Ed.), Models in Paleobiology, 82-115, San Francisco: Freeman, Cooper & Co.

9. Empirical Test, Academy of Management Journal, 37(5), 1141-1166.

10. eMarketer. "Worldwide Smartphone Usage to Grow 25% in 2014." 11 June 2014. eMarketer. 1 November 2014 <http://www.emarketer.com/Article/Worldwide-Smartphone-Usage-Grow-25-2014/1010920>.

11. Forester, T. (1989). Computers in the human content. Cambridge, Ma: MIT Press

12. Frankston, R. (1974). THE COMPUTER UTILITY AS A MARKETPLACE FOR COMPUTER SERVICES http://www.frankston.com/public/?name=TR128

13. Gartner (2010). Hype chart retrieved from http://www.gartner.com/it/page.jsp?id=1447613 May 12, 2011

14. Gharajedaghi, J., (2006). *System Thinking: Managing Chaos and complexity: A platform for designing business architecture.* Butterworth-Heinemann, Boston, Oxford

15. Giere, R. (1999), Science without laws. Chicago, IL: University of Chicago Press.

16. Golden, B. (2011). The cloud CIO: A tale of two IT futures. Retrieved May 2, 2011 from http://www.cio.

17. Horrigan, J. A., (2008, September). Use of cloud computing applications and services. *Pew Internet and American Life Project.*

18. IBM 1620, (2011). IBM 1620 Retrieved May 2, 2011, from http://www.ebroadcast.com.au/lookup/encyclopedi a/ib/IBM_1620.html

19. IBM (2011). Highlights retrieved May 2, 2011 from
    a. http://www-03.ibm.com/ibm/history/documents/pdf/188 5-1969.pdf

20. IDC (2010). Cloud Computing in the Midmarket : Assessing the Options Adapted from *Worldwide Small and Medium-Sized Business 2010–2014 Forecast: Recovery and Change in SMB IT Spending by Category and Region* by Ray Boggs, IDC #222409

21. Knorr, E. (2011). Cloud computing: IT as a commodity. Retrieved May 2, 2011, from http://www.infoworld.com/print/149913

22. Kranzberg, M. (1989), IT as revolution: The information age. In T. Forester (Ed.). Computers in the human context (pp.19-32). Cambridge, MA: MIT Press.

23. Kuhn, T.S. (1996), Structure of scientific revolutions. Chicago: University of Chicago Press.

24. Lucas, H. C. (2008). *Inside the future: Surviving the technology revolution.* Connecticut: Praeger

25. Lynch, C. G. (2009, March 12) How and why to Launch a business presence on twitter retrieved May 12, 2008 from http://www.cio.com/article/484266/

26. Martz, B., & Cata, T. (2008, November December). Students' Perception of IS Academic Programs, IS Careers, and Outsourcing. Journal of Education for business. 118-125

27. Minsky, M. (1988). *The Society of Mind.* New York: Simon & Schuster.

28. Moser, M. (2009). Workload automation: Helping cloud computing take flight. Retrieved May 1, 2011, from http://documents.bmc.com/products/documents/39/17/123917/123917.pdf

29. Moore, G. E. (1965). Cramming more components onto integrated circuits. *Electronics Magazine, 38(8).*

30. Mumford, L. (1970), Myth of the machine: The pentagon of power. New York: Harcourt.

31. Ozzie, R. (2008). After Bill. *The Economist*, June 28, 2008, p. 77.

32. Ponschock, R. L. (2007). Computer technology, digital transactions, and legal discovery: A phenomenological study of possible paradoxes (Doctoral dissertation, Capella University, 2007) (UMI No. 3246872).

33. Prahalad, C.K. & Krishnan, M.S. (2008). *The new age of innovation: Driving co-created value through global networks.* New York: McGraw-Hill.

34. Robey, D. (1981, October). Computer information Systems and Organization Structure *Communication of the ACM.* October 1981 v.24 Number 10

35. Rogers, E.M. (1965). *Diffusion of Innovations*, 4th edn. The free Press, New York

36. Romanelli, E., & Tushman, M. "Organizational Transformation as Punctuated Equilibrium: An Empirical Test." *Academy of Management Journal* 37 (1994): 1141-1166.

37. Sambamurthy V., Bharadwaj, A., & Grover V. (2003). Shaping agility through digital options: Reconceptualizing the Role of Information Technology in Contemporary Firms. *MIS Quarterly* Vol. 27, No. 2 (Jun., 2003), pp. 237-263

38. Social Networking to Academic Networking ... A PARADIGM SHIFT. By: Childers, Tim, Internet@Schools, 2156843X, May/Jun2011, Vol. 18, Issue 3

39. Stefanic, A. (2004). The business case for outsourcing. Employee Benefit Plan Review, 58(8), 11-13.

40. Tanoglu I., Basoglu N., & Daim, T. (2010). Exploring Technology diffusion: Case of Information *Technologies International Journal of Information Technology and Decision Making*
    a. Vol 9, no 2 195-222

41. Trauth, E. M., Farwell, D. W., & Lee, D. (1993). The IS expectation gap: Industry expectations versus academic preparation. *MIS Quarterly,17*, 293–307.

42. Waskow, R. A. (2004). Free time/free people statement: Freeing our time. Retrieved November 14, 2006, from http://www.shalomctr.org/taxonomy/term/45

43. Childers, A. (2011, Jan). Niche social networks in a Facebook era. Retrieved May 3, 2012, from http://www.adrianchilders.com/niche-social-networks-in-a-facebook-era/

44. Harrington, L. (2007, April). Defining technology trends. Inbound Logistics. Retrieved May 15, 2008, from http://www.inboundlogistics.com/articles/features/0407_feature01.shtml

45. Bell, G., Cady, R., McFarland, B., Delagi, J., O'Laughlin, Noonan, R., & Wulf, W. (1970). A new architecture for mini-computers- DEC PDP-11. *AFIPS Conference Proceedings*, Vol. 36 pp. 657-675.

46. Purcell, K., Rainie, L., Rosenstiel, T., & Mitchell, A. (2014). How mobile devices are changing community information environments - Part 1: Mobile news takes off. Retrieved March 12, 2015

from http://www.pewInternet.org/2011/03/14/part-1-mobile-news-takes-off/

47. Anderson, J., Q., & Raineie, L. (2010). The future of social relations. Retrieved March 20, 2015 from http://www.pewInternet.org/files/old-media/Files/Reports/2010/PIP_Future_of_Internet_%202010_social_relations.pdf

48. Turner, M., & Gens, F., (2009). Cloud computing drives breakthrough improvements in IT service delivery, speed, and costs. Retrieved March 20, 2015 from http://www.304.ibm.com/industries/publicsector/fileserve?contentid=250743

49. Ward, J., & Peppard, J., (2009). *Strategic planning for information systems*. Bedfordshire, UK: Wiley.

Chapter 3: Virtualization a Global Reality

Virtualization or the creation of a simulated rather than actual reality has created a heightened impasse for members of the 21st century society with a key focus on transformative change. Driven by technological innovations, virtualization has instigated an evolution, as well as a revolution, on the societal landscape. Virtualization impacts society, the economy, and the heart of organizations at the core of their operating culture. Citizens of the 21st century must have the knowledge and skills necessary to navigate through these complex uncharted waters.

Dwellers of the 21st century must understand the rapid change dynamic that exists in digital environments and readily adapt.

Background

General Motors launched a marketing campaign around the slogan "*This is not your father's Oldsmobile*". If one considers that slogan in relation to today's rapidly changing technological environment, correspondingly "*this is not your father's society*". Technology and the Internet have initiated a plethora of changes deeply influencing our very existence -- from the individual to global corporations and beyond. Sociological participants of the 21st century are tasked to navigate uncharted waters including technology that increasingly drives the family automobile and the overall cyber universe. Individuals are being continually challenged with protecting their privacy in ways not imagined even a decade ago, while retaining and protecting their personal identity. The message of this book is one of enlightenment. *Will the path that we are on create a new individual of humanity? Are we as a species and a society evolving into a DIGIPERSON?* Society is something that encompasses the very processes of life itself.[11] The Internet and technology at-large, with its absence of geographic boundaries, are requiring different skills and different thinking from those who have led transformative change in the past. In today's rapidly changing technological atmosphere, change and complexity are the norm, and the status quo is no longer preset or finite. The new constant in business is *change*.

In the 21st century, transformative change is enmeshed in a web where organizations and knowledge workers are networked in an economy that thrives in an ever increasingly turbulent world.[8] As Kelly suggested, "the dynamic of our society, and particularly our new economy, will increasingly obey the logic of the networks".[22] By its very nature, *transformative change* implies that the system, or people that comprise the system, cannot return to the old way of doing things; otherwise, the change is not truly transformative.

> *Sociology, being especially concerned with social evolution, has a new and distinct factor at work which we may call association, cooperation or combination, and this it is which gives sociology its distinct place in the list of general sciences. (Ellwood, 1911)* [11]

As Ackoff cautioned, answers to the future are not always encapsulated in the past because the future continues to write and rewrite its story. [1] Leading others to *transformative change* in global enterprises relies on a new paradigm of virtualization operating at warp speed.[25] In many cases, *virtuality* may be as overwhelming and intimidating as complexity[6, 9, 5, 16] and become so dramatic that in many instances it may require organizations to reassess and possibly redesign its very core and culture.[43]

A Fatal Retraction

It was just noted that total transformation by definition concludes that there is no going back to an existence of the past. What if the science that enabled the societal digitization to progress to its current state is the same science that could sever our digital luxuries? The following may sound like science fiction and also play into what some believe to be fringe psychosis. The current term "preppers" describe

54

individuals who want to live *"off the grid"* in fear of an event that could force us back into an era without digital electronics. One could argue that the 21st century society has already made the transformation into digital reliance. The sociological participants of the 21st century have made the transformation to a digital-reliant lifestyle and as a society cannot revert back. Consider however a nuclear pulse, that goes by the acronym EMP; a short burst of electromagnetic energy is deployed miles above the atmosphere. This event could drive society back to the 1800's. Based on society's engagement with electronic devices, societal participants rely heavily on *biotechnics* (i.e. hearing aids, pacemakers, and oxygen generators); consider that the sudden absence of these devices might result in sudden death for some, and severe immobility issues for others.

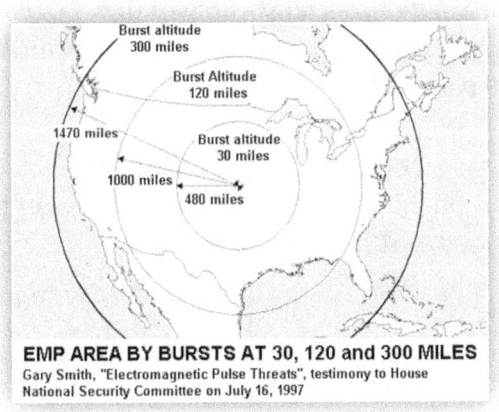

EMP AREA BY BURSTS AT 30, 120 and 300 MILES
Gary Smith, "Electromagnetic Pulse Threats", testimony to House National Security Committee on July 16, 1997

Figure 8: EMP Bursts

Transitioning to a broader view, the sudden and protracted failure of the total power grid would result in a loss of the

55

communication network, transportation, and a stagnation of the total economy. In a very short period of time millions of lives could be lost because of starvation, riots, and personal hopelessness.

This might harken towards science fiction; however, Former CIA Director James Woolsey had this to say about North Korea launching an EMP attack against the United States in a statement presented to the House Armed Services Committee:

> *There is now an increasing likelihood that rogue nations such as North Korea, and before long, most likely, Iran, will soon match Russia and China in that they will have the primary ingredients for an EMP attack. ... Two thirds of the US population would likely perish from starvation, disease, and societal breakdown. Other experts estimate the likely loss to be closer to 90 percent.*[33]

The U.S. is vulnerable[19] and "going back" could cause destruction on an immeasurable level. Although EMP has been analyzed, written about and

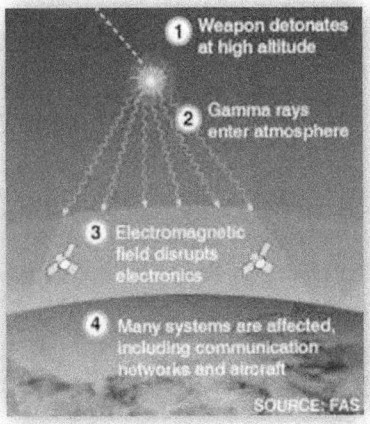

Figure 9: EMP

testified on to congress, as a nation and a global society minimal effort has been expended to prevent the devastation and retraction that could take place if an event should occur.[30] It is currently an "Achilles heel".

Virtualization of Society

Why is virtualization and its effect on society so important and influential? Since inventor Benjamin Franklin's infamous lightening rod and kite demonstrated the power of conductivity, people have been intrigued by the invisible force and power of electricity. Some have been mystified and others intimidated. In the modern world, consumers have been using electronic devices for the greater part of their lives, often taking their existence for granted.[3] It has been suggested that electronic technology has influenced society in ways that may prove to have a greater impact than the Industrial Revolution.[27]

A century ago, social interactions involved relationships with others who were within a short walking distance.[12] For many, especially in the industrialized West, small face-to-face communities are rapidly disappearing. Technology and the Internet have introduced communities that do not geographically exist and have no tangible physical presence. These virtual villages or townships[34] are not represented by geography, social class, or financial accounting. Rather, their cyber position is defined and driven by curiosity.[26] Seminal sociologist Charles Ellwood opines: "The facts of contemporary social life are indeed even more important to the sociologist that the facts of history.[11] As Laurie Anderson musician/artist wrote "Technology is the campfire around which we gather". The legacies of 100 million virtual socialization network subscribers are now being buried in the form of personal likes, dislikes, dreams, and possibly "dirty laundry" in landfills of virtual villages or virtual communities like FACEBOOK, Twitter, Instagram, and Snapchat.[7]

Dating and matchmaking have taken on not only a digital environment, but have become an increasing component of an entire industry. In the not so distant past, an individual looking for a social companion as a teen would go to a school dance or even a retail mall. As an adult, a lounge or bar might be the place to socialize and find companionship. The 21st century has added a new means of finding that special someone. Online dating websites now are available in many specialties or niche focuses. There is Christianmingle.com; Jdate.com for Jewish singles, and even Farmersonly.com. There are numerous sites in each of these categories. In the rural category there are even subcategories like Countrymatch.com. There is one website

Top10bestdatingsites.com that ranks the popularity of existing sites!

The concept of *virtual villages* is now being adopted by large corporations seeking community-building across networks. Companies are using social networking services like Facebook, allowing employees to upload photos of themselves—not just corporate ID mug-shots, as well as personal information. The network is easily accessed by other employees giving large companies a sense of community. Through the network, members are also able to search for best practices on marketing or sales topics, dialogue through blogs and wikis written by informed employees, along with providing instant entrée to an in-house encyclopedia considered *social networks*. Conceptually, it will boost top and bottom lines by providing clearer understandings of who knows what and who knows whom, within a company and among its business partners. In essence, these types of networks have the capacity to operate as central nervous systems[29] for organizations becoming more valuable as they are trained and tested.

Leadership of the 21st century will need to move people and organizations from the prototypical brick and mortar landscape towards virtual borderless thinking. With the proliferations of enterprise networks and new technologies, operating beyond traditional organizational charts and outside traditional management hierarchies will require transformative change using widely distributed multi-minded schemas and expertise.[14] As Gharajedaghi & Ackoff [15] suggested, this type of understanding allows for creativity and design where the future can be co-created to foster more than just information and knowledge.[37] Instead, through transformative change efforts, organizational leadership interacting with others can cooperatively thrive by understanding new technologies and their impact on organizations, society and the future.

Aligned with these types of transformations, Deutschman in his study of change, reported three major factors involved in making a positive difference in increasing people's mastery and the ability to change: reframing change, radical change, and multifaceted support of change. He found that the way change is framed is important. The reframed message must be positive and inspiring.[10] According to Howard Gardner, simplicity of the reframed message, embraced with emotion, typically results in a positive, longer lasting outcome. [10] How change propositions are framed can create angst or inspire and sustain them.[17] Essentially mastering transformative change is necessary because time and resources are too limited to keep revising transformational change processes without success, especially given the rapidity of technological innovation.

Virtualization and the Economy

What impact does virtualization have on the economy? In the networked digital economy, technology and computer driven communication have had a dramatic effect on the traditional industrial distribution value chain.[35] Prior to digital economy, sources of goods and services attempted to stay close to the consumer.[41] In the Internet environment, there are fewer intermediaries. Kelly furthered this thought with the statement "the network economy shifts places to spaces".[21] As an Internet customer, consumers are creating a pull model[28] by scouring the global landscape and taking charge of their purchases and decisions. Kelly submitted that in the digital purchasing age, value flows from abundance. Abundance and plentitude now govern the network economy where duplication and replication drive values, works to open systems, and creates opportunity as barriers to entry evaporate.[21]

Many conventional companies have approached the digital era as a new medium to extend traditional business through an additional distribution channel. In contrast, leading companies in the digital landscape approached the digital environment organically[15] with new thinking and new actions embracing transformational change through network ecology's of organisms[45, 46] that are constantly in flux.[22] As the future technology of radio frequency identification chips (RFID), Smart Cards, Personal Data Assistants (PDA), Bio and Nano-technology continues to advance exponentially, organizational leadership needs to be poised to seek opportunities by embracing transformative change and innovation in the digital economy.

Virtualization's Impact on Employees

How do organizations deal with virtual employees? The boundaries between home and workplace, actually anywhere and workplace have become blurred in recent years.[13] As computer networks have incorporated fiber optics, cable modems, DSL, and satellite into a digital matrix, cyberspace offers more and more possibilities for off-premise work – commonly referred to as telecommuting.[3] Broadband capabilities to the home, wireless networks within the house, and software such as virtual private networks (VPN) or the use of smart phone "apps", now enable users to gain access to their company's computer from remote locations. These technology options contribute to a transition and integration of office activities which further influence employees' personal lives. Friedman[13] concluded that personal home computers have added dimensions visually and interactively, creating homes that now operate as school and workplace. Yet, people are by their very nature social animals requiring interaction[2] and multiple sources of stimulation (visual, audio, tactile). Vance accentuated the paradox of these basic instinctual needs, by predicting a time when most people will work from home without needing to visit an office.[43]

Another distinguishing attribute of the true technological reconstruction is that as this technology paradigm shifts [24] many workplace innovations are occurring simultaneously. [23] Although, working from home has tremendous advantages, it comes with challenges as well. One of the biggest challenges with working from home is not being able to switch off the responsibility when walking out of an office at the end of the day.[32] This phenomenon of fuzzy work boundaries[30] requires employees to balance personal needs and leisure activities, adding to the complexity and demands for worker productivity.

What is new in the evolving organizational landscape is the increased assimilation of a great deal of an employee's private life and activities into systems that are supplied on a computer desktop. In the past, telephone and company mail were tools an employee may have used to conduct personal business. Today, employees have access to the Internet, cell phones, personal digital assistants (PDA), and laptop computers. This cadre introduces a remarkable new range of potential personal uses that may legitimately interfere with productivity, or even expose the employer to legal action or financial liability.[42, 47]

The intersection of an employee's personal time and work life has become blurred for employees and employers. Work now is frequently performed at home or other off-premise locations. The work may be performed on a computer or other electronic appliances supplied by the company or a system that is owned by the employee. The question of who owns the property rights to the work completed on the employee's personal computer[31] is sometimes unclear. If the work product was created on the individual's own time, with their own computer, in their own home, the answer should be clear, should it not? Yet, the employer may expect, or assume, that the organization has intellectual property rights to the contents related to work tasks while the employee may have thoughts on the other side of this paradoxical quandary.

Employees have become fast-paced and time-constrained. According to Waskow, Americans now work harder and longer, in accord with other people's schedules, than in prior decades.[44] A shift is also occurring in the way members of our technologically oriented society interact. The virtual employee's work life is shifting from one of personal relationships between individuals, to one with a more virtual bond. As technology has increased over the past 10 years, many close friends may have never met face-to-face. Balancing this conundrum where work life has been seen as being part of everyday life, and work has been socially constructed,[4] requires understanding human interactions. It is important for people to experience a sense of connection [38] and accomplishment for[40] as well as self-efficacy to be activated by employees in an organic goal-orientated productive fashion. Importantly, the motivational component creates a sense of hope through agency thinking or goal-directed thought. This future transformative thinking necessitates both the perceived capacity to envision workable actions and goal-directed energy[40] necessary to attain the actions. Innovation comes about when people are enabled to use their full brains and abilities instead of being constrained. By unleashing positive emotions which flow from perceptions of successful goal pursuits, self-esteem is enhanced.[20] When people feel valued and intrinsically motivated they are fully engaged in their work which in turn propels them to unleash their human potential.

Balancing Technological Complexity with Organizational Sustainability

Why is it important to balance technological complexity with organizational sustainability? Martin posited that the "flux is good",[28] indicating that in the Internet landscape tweaking change will be the norm. This is an environment where the ink never dries. Managers and leaders working within the digital landscape need to understand this virtual dynamic. In the virtual environment, sustainability may not be measured in decades as it once was in the industrial period. Sustainability requires a multi-minded systems view. By asking questions, with a commitment to transformative change, by embracing technology and attendant influences, managers and leaders can dramatically influence how organizations bridge technological complexity and organizational sustainability.[14]

Social networks integrating employees, customers, and suppliers were not even contemplated in the not too distant past. Yet, if today's leaders and managers overlook these virtual connections, they will be doing a disservice to their organizations. As Prahalad & Krishnan noted, "There is a fundamental transformation of business underway. Forged by digitization, ubiquitous connectivity, and globalization, this will radically alter the very nature of the firm and how it creates value...".[36] Partnerships between customers and suppliers are morphing into global, complex, interdependent exchange points, forcing organizations to extend planning beyond the four walls of their enterprise.[18] Many firms find themselves unprepared to accept the challenges posed by this new reality as managers and leaders face deeply ingrained organizational legacies steeped in both social and technological agendas.[36] Management in the 21st century plays a pivotal role in transforming mindsets, skills, behaviors, and decision structures of organizational leadership. This transformation to a virtual social architecture will only be successful through a period of socialization and acculturation where the organization and its component parts work together organically to effectuate transformative change.

Given the various elemental digitization events that have transpired and continue to gain velocity, it appears evident that a holistic interrelationship between the global sectors is upon us, especially related to the business, academic and political landscapes. Harrington posited that these technological advances will continue at an accelerating pace, thereby inhibiting individual leverage.[18]

Consider that theoretical constructs dating 20-30 years were a forbearer of these technological innovations and convergence. Beniger discussed these very advances as having the potential for unique leverage as well as a societal convergence.[48] Consider that the movie "Back to the Future" portrayed many technological innovations back in the late 1980's that have come to fruition. Therefore, the technological advances are a speciation event formed and evolving to further the convergence of global sectors resulting in sustainable and positive leverage.

Chapter Wrap-up

The digitization of life in the 21st century has infiltrated our very existence, only a few of which were examined herein. The transformation events in the 21st century related to social networking and the *CLOUD* is both the ends and the means. Social networking is a powerful force throughout the world. Consider the power of Twitter and other microblogging tools in political events in Egypt, Libya, North Korea, etc. The entire world uses these tools everywhere. Childers added "Twitter and Facebook are fantastic at what they do. Connecting you with your friends, but look forward to the next generation of niche social networking sites, centered on every type of interest imaginable."[53]

Beniger, a seminal thinker on society and evolutionary control, posited:

> *As in earlier revolutions in matter and energy technologies, the nineteenth-century revolution in information technology was predicated on, if not directly caused by, social changes associated with earlier innovations"; Because technology defines the limits on what a society can do, technological*

innovation might be expected to be a major impetus to social change.[48]

Social networking enabled by the Internet is performing crucial roles throughout society. Consider a few positive purposes that social networks provide:

- Researchers can collaborate from any location around the world
- Friends/families are no longer bound by distance
- Academics use social networks to enhance courses and knowledge transfer
- Businesses use social networks internally and along the supply chain

Continual digital-only interaction can actually alter brain circuitry. "The brains of the younger generation are digitally hardwired from toddlerhood, often at the expense of neural circuitry that controls one-on-one people skills."[54]

Digitization and social networking have migrated into businesses of all sizes.[52] Outside of the boundaries of industry, 69% of online Americans use webmail services, store data online, or use software programs such as word processing applications whose functionality are located on a web based service provider. In doing so, these users are making use of "cloud computing," an emerging architecture by which data and applications reside in cyberspace, allowing users to access them through any web-connected device.[49]

Through social constructionism, individual behaviors and group participation in the building of perceived social realism can be uncovered. Digitization of society and "Internet-based, just-in-time delivery of data, applications, storage, and computing power as services are done in a way that completely shields consumers from underlying technical details".[50] *Digitization* in general and *Social Networking* have profoundly changed and continually fuels social change. On traditional Darwinian account of organic evolution, variation is supplied primarily by mutations which are random relative to the potential beneficial contribution to fitness of the modified trait.[50] "Evolutionary theory applies to *populations* of individuals".[51]

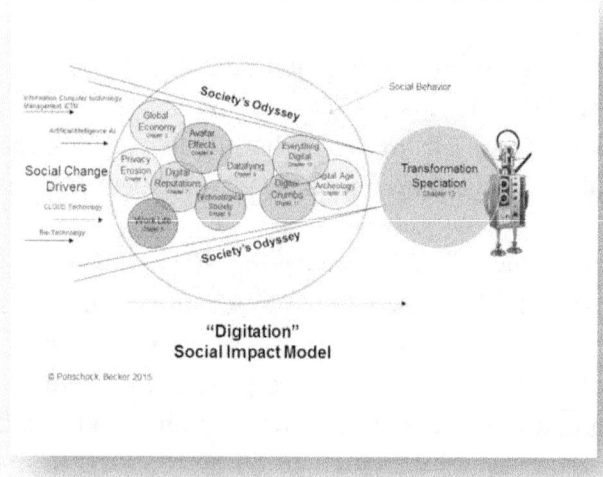

Figure 10: Digitization Social Impact Model

The evolution of new species doesn't usually happen in an afternoon, so it's quite difficult to see. And, on the geological scale, mankind has been around for a blink of an eye. Still, we have observed speciation and see its telltale fingerprints everywhere. Evolutionary theory is a very intellectually-powerful concept indeed. It has pushed biology to the fore of science and it has given answers to questions that would have otherwise gone unanswered or simply put down to an infamous deity's 'mysterious ways'. Mankind has always striven to understand life, the universe and everything; the theory of evolution goes a long, long way to reach this lofty, ambitious goal.[51]

The relationships depicted in the "convergence model" illustrate the junction of previously bounded and diverse facets of societal life becoming a single new entity. The domains previously held beyond arm's length are melded into a digital environment creating an atmosphere that will permanently change society and the way humans interact.

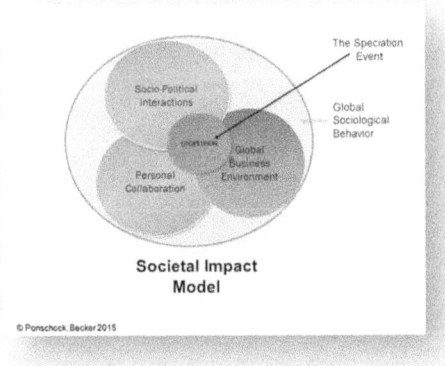

Figure 11: Convergence Model

Civilized society is made of laws. In developed nations, laws attempt to protect the privacy of its citizens and with privacy there are barriers in place to avoid government from just stumping on our inalienable rights. The next chapter discusses how digitization places these rights in harm's way.

End Notes

1. Ackoff, R.L. (1974). *Redesigning the future.* New York: John Wiley & Sons.
2. Aronson, E. (1992). *The social animal* (6th ed.). New York: W.H. Freeman.
3. Barnatt, C. (1995). Office space, cyberspace, & virtual organization. *Journal of General Management,20*(4), 78-91.
4. Berger, P.L., & Luckman, T. (1967). *The social construction of reality: A treatise in the sociology of knowledge,* Garden City, NY: Doubleday.
5. Briggs, J., & Peat, F. D. (1990). *Turbulent mirror: An illustrated guide to chaos theory and the science of wholeness.* New York: Harper & Row.
6. Brown, S. L., & Eisenhardt, K. M. (1997). The art of continuous change: Linking complexity theory and time-paced evolution in relentlessly shifting organizations. *Administrative Science Quarterly, 42*(1), 1–34.
7. Cashmore, P. (2006, August, 9). MySpace hits 100 million accounts: Mashable social networking 2.0. Retrieved November 9, 2006, from http://mashable.com/2006/08/09/ myspace-hits-100-million-accounts/
8. Capra, F. (2002). *The hidden connections: Integrating the biological, cognitive, and social dimensions of life into sustainability.* New York: Doubleday.
9. Csikszentmihalyi, M. (1990). *Flow: The psychology of optimal experience.* New York: Harper Perennial.
10. Deutschman, A. (2005). Make change or die. *Fast Company,* 94(5), 52-60.

11. Ellwood, C., (1911). Society and modern social problems. Retrieved March 20, 2015 from http://www.archive.org/stream/cu31924013932128/ cu31924013932128_djvu.txt

12. Ermann, M. D., Williams, M. B., & Shauf, M. S. (1997). *Computers, ethics and society*. New York: Oxford University Press.

13. Friedman, T. L. (2005). *The world is flat: A brief history of the twenty-first century*. New York: Farrar, Strauss, & Giroux.

14. Gharajedaghi, J. (2006). *Systems Thinking: Managing chaos and complexity: A platform for designing business architecture*. Oxford, UK: Elsevier.

15. Gharajedaghi, J., & Ackoff, R. L. (1984). Mechanisms, organisms and social systems. *Strategic Management Journal, 3*(5), 289-300.

16. Gleick, J. (1987). *Chaos: Making a new science*. New York: Penguin Books.

17. Hammond, J.S., Keeney, R.L. & Raiffa, H. (1998). The Perfect decision. *Inc., 20*(14) 74-78.

18. Harrington, L. (2007, April). Defining technology trends *Inbound Logistics*. Retrieved May 15, 2008 from http://www.inboundlogistics.com/articles/features/0407_feature01.shtml

19. Harrison, R. (2014, January 28). Fatal inaction: The U.S. remains vulnerable to EMP attacks. US News and World Report retrieved September 29, 2014 from http://www.usnews.com/opinion/blogs/world-report/2014/01/28/failing-to-guard-against-electromagnetic-pulse-attacks-could-be-fatal

20. Hewitt, J. P. (1998). *The myth of self-esteem: Finding happiness and solving problems in America.* New York: St. Martin's Press.
21. Kelly, K. (1999). New *rules for the new economy: 10 radical strategies for a connected world.* New York: Penguin Books.
22. Kelly, K. (1997). New rules of the new economy: Twelve dependable principles for thriving in a turbulent world. *Wired, 5*(9) 140-143, 186-197.
23. Kranzberg, M. (1989). IT as revolution: The information age. In T. Forester (Ed.). *Computers in the human context* (pp. 19-32). Cambridge, MA: MIT Press.
24. Kuhn, T.S. (1996). *Structure of scientific revolutions.* Chicago: University of Chicago Press.
25. Loehr, J., & Schwartz, T. (2003). *The power of full engagement.* New York: Free Press.
26. Luthra, N. (2006). The "Real" and the "Virtual" in public space (Master Thesis, University of New York at Buffalo). (UMI 1431955).
27. Manzano, Y. (1999). Technology's influence in 20th century life and its difference from industrialization's Influence in 19th century life. Retrieved January 3, 2006,   from http://ww2.csf.fsu.edu/~manzano/writing/essays/history/technology/html
28. Martin, C. (1996). *The digital estate: Strategies for competing, surviving, and thriving in the Internet worked world.* New York: Mc Graw-Hill.
29. Minsky, M. (1988). *The society of mind.* New York: Simon & Schuster.
30. McNeill, D. & Freiberger, P. (1993). *Fuzzy logic: The revolutionary computer technology that is changing*

our world. New York: Simon & Schuster.

31. Montana, J. C. (2005). Who owns business data on personally owned computers. *The Information Management Journal. 39*(3), 36-42.

32. Montero, P. (2004). Two perspectives on how to work from anywhere. *The Journal for Quality and Participation, 27*(3), 24-28.

33. Mulrine, A. (2014, August). Is US vulnerable to EMP attack? A doomsday warning, and its skeptics. *Christian Science Monitor,* p.N.PAG.

34. Ponschock, R., & Greif, T. B. (2007). Archeological excavating in virtual villages: A primer on discovery of artifacts from a digital community. *Proceedings of the IABE 2007 Annual Conference, 3* (1), 260-265.

35. Porter, M.E. (1985). *Competitive Advantage.* New York: Free Press.

36. Prahalad, C.K. & Krishnan, M.S. (2008). *The new age of innovation: Driving co-created value through global networks.* New York: McGraw-Hill.

37. Prahalad, C. K., & Ramaswamy, V. (2004). *The future of competition: Co-creating unique value with the customers.* Cambridge, MA: Harvard Business Press.

38. Rheingold, H. (2000). *The virtual community: Homesteading on the electronic frontier.* Boston: MIT Press.

39. Snyder, C. R., Irving, L. & Anderson, J. R. (1991). Hope and health: Measuring the will and the ways. In C. R. Synder & D. R. Forsyth (Eds.), *Handbook of social and clinical psychology: The health perspective* (pp. 285-305). Elmsford, NY: Pergamon.

40. Snyder, C. R., Rand, K. L., & Sigmon, D. R. (2005). Hope theory: A member of the positive psychology

family. In C. R. Snyder & S. J. Lopez (Eds.), *Handbook of positive psychology* (pp. 257-276). New York: Oxford University Press.

41. Thompson, A. III, Strickland, A. J. & Gamble. J. (2006). Crafting and Executive Strategy (15th ed.). New York: McGraw-Hill.

42. Townsend, A. M., Alberts, R. J. & Whitman, M. E. (2000, March) Employer liability under the communications decency act: Developing effective policy response. *Employee Responsibilities and Rights Journal*, (3)39-46.

43. Vance, M., & Deacon, D. (1995). Think out of the box. Franklin Lakes, NJ: Career Press.

44. Waskow, R., A. (2004). *Free time/free people statement: Freeing our time.* Retrieved November 14, 2006, from http://www.shalomctr.org/taxonomy/term/45

45. Wheatley, M. J. (1999). *Leadership and the new science.* San Francisco: Berrett-Koehler.

46. Wheatley, M. J. (2005). *Finding our way: Leadership for an uncertain time.* San Francisco: Berrett/Koehler.

47. Whitman, M. E., Townsend, A.M., & Alberts, R.J. (1999, June). The communications decency act is not as dead as you think. *Communications of the ACM*, 42, 15-17. personal psychology. drleestjohn@earthlink.net

48. Beniger, J. R. (1986). *The control revolution: Technological and economic origins of the information society.* Cambridge, MA, Harvard University Press.

49. Horrigan, J., (2008) Use of Cloud Computing Applications and Services. Retrieved March 23, 2015 from http://www.pewInternet.org/2008/09/12/use-of-cloud-computing-applications-and-services/

50. Moser, M. (2009). Workload automation: Helping cloud computing take flight. Retrieved May 1, 2011, from http://documents.bmc.com/products/documents/39/17/123917/123917.pdf

51. Evolution (2002). The theory of evolution – part II. Retrieved May 1, 2012, from http://h2g2.com/dna/h2g2/alabaster/A737985

52. IDC, (2010). Cloud Computing in the Midmarket : Assessing the Options Adapted from *Worldwide Small and Medium-Sized Business 2010–2014 Forecast: Recovery and Change in SMB IT Spending by Category and Region* by Ray Boggs, IDC #222409

53. Childers, A., (2011, Jan). Niche social networks in a Facebook era. Retrieved May 3, 2012, from http://www.adrianchilders.com/niche-social-networks-in-a-facebook-era/

54. Small, G., and Vorgan, G., (2008). *iBrain: Surviving the technological alteration of the modern mind*. Harper Collins Publishers. NewYork, NY.

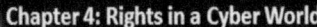

Chapter 4: Rights in a Cyber World

Societal transformation to a networked digital society driven by ubiquitous computers and "CLOUD" architecture may also be attacking the very roots of our personal being - privacy. Privacy battles have only begun. In 2003 and 2004, privacy battles began to rage because of peer-to-peer networks that were downloading and copying music via the Internet. The tracking of these and other transactions has led directly into the homes of private citizens. [34] *As soon as a way to close one leak of private information is found, a new weakness through which to disclose personal data is discovered.* [41]

Challenges through newer technology, such as radio frequency identification chips (RFID), are being formed. [4] *Understanding the evolution of computer technology and the speed at which it is changing is an important basis for understanding the difficulties facing law enforcement and the judiciary when digitally encoded personal data is involved.*

Background

The maxim "every man's home is his castle" is a highly respected principle that was enshrined in the Semayne case of 1603. Some historians indicate that the "knock–and–announce" rule was grounded in the Semayne case. Others trace it as far back as 1275 to the English common law protection of the home as the castle of defense and asylum. The English court recognized the right of the homeowner to defend his house against unlawful entry by the King's agents. At the same time, the authority of law enforcement to break and enter upon proper notice by appropriate officers was acknowledged.[37]

The Fourth Amendment of the Constitution of the United States was adopted by congress in 1789 and ratified by the states in 1791. It too included language that protected the citizenry from illegal search and seizure.

The right of the people to be secure in their persons, houses, papers, and effects, against unreasonable searches and seizures, shall not be violated, and no Warrants shall issue, but upon probable cause, supported by Oath or affirmation, and particularly describing the place to searched, and the persons or things to be seized. [47]

Applying the language of our forefathers to life in the 21st century is now in the hands and minds of today's judiciary. The post-September-11th Patriot Act was the latest in Fourth Amendment acts to impact legislation. This act has given government latitude, especially in surveillance and access into the private lives of every American. This Act has stirred rights advocacy groups, legislators, and supporters alike.

The phenomenon of cyberspace and its medium, the Internet, recently have generated an abundance of debate in legal literature. The dialog attempts to answer the question of whether traditional doctrine is capable of handling the legal issues generated by the Internet and the current digital landscape.[19]

Rights and Principles in a Cyber Era

As discussed in a previous chapter, technology has become embedded into every aspect of our very existence and its presence approaches invisibility; our daily lives and personal privacy may be pressured or even partially surrendered.[23] For example, retailers led by Wal-Mart are consorting with major companies like IBM, NCR, Gillette, and Procter & Gamble to develop a future road map that calls for monitoring and tracking individuals and using collected information for commercial use. As the development of information appliances accelerates, so could the opposition to its implementation. Opponents to RFID argue that the plans integrate and embed RFID tags in products and loyalty cards.[4] The retail industry is facing opposition from consumer action groups. In his seminal 1991 *Scientific American* article "The Computer for the 21st Century," Mark Weiser, an early visionary, surmised that ubiquitous computing is more a social control issue than one of privacy.[52] Today's concern, an ongoing public debate over RFID technology, continues as it relates to consumer data privacy in the retail industry, where the tension between control and privacy is evident.

> *The trails of personal information left in a person's wake are nearly endless—emissions ranging from the DNA in a skin flake to the digital exhaust willingly or unknowingly produced by cell phones, computers, and other technologies. How that personal information is collected and used by a third party is a concern for leaders and legal scholars in the policing community.* [36]

The concern over leaking digitally encoded personal data has stirred headlines from many privacy action groups. As individuals browse through web sites, parties on the other end of those sites may be browsing through their personal information and using that information for their own purposes. [8] The Privacy Rights Clearing House estimated in a posting last updated November 4, 2005, that more than 51 million United States citizens may have had their personal data compromised since February 2005.[15]

Warrants

Knock and Announce

Throughout the course of history and in today's technologically advanced world, search and seizure rules have continued to play an important role in democratic societies. The processes used by law enforcement are the root of judicial debate and action. The "knock-and-announce rule" requires police officers serving an arrest or search warrant to knock and announce their authority and

80

purpose. Police officers are also required to wait for a reasonable period of time before forcibly entering a home or office. Law enforcement cannot simply walk into a person's home with a subpoena. Standard procedure requires the officer to knock and state, "Police; search warrant." Before the Nebraska Supreme Court in *Lammers v. Nebraska* (2004) Lammers argued that the officers did not provide adequate time in the execution of the knock-and-announce rule before breaking down the door. In testimony, the officers indicated they first yelled "police department search warrant" and waited 8 seconds and then did a second knock and announce. After waiting another 10 to 15 seconds, the officer then broke down the door. The *Lammers v. Nebraska* court concluded when it weighed the totality of evidence that reasonable suspicion of exigency was present, and the forcible entry was allowed under the law. [30, 55]

No-knock Exclusion

There are special circumstances when the knock-and-announce rule may be set aside. "No-knock" warrants can be issued as exclusionary procedures. State courts, as well as the Supreme Court of the United States, have written decisions on several aspects of the knock-and-announce rule and the section of the Fourth Amendment to the Constitution. In 1995, the Supreme Court held that no-knock warrants violate the Fourth Amendment; however, under compelling conditions, such as strong belief that the occupant may be armed or that the evidence being sought could be easily destroyed, a no-knock entry may be legally reasonable.[11]

81

This rule has been analyzed and debated in the lower courts and the Supreme Court on numerous occasions. The court was split in the case of *Ker v. California* (1963). Without securing a search warrant, the officers obtained a passkey from the building manager. One of the officers had information that the defendant's husband was selling marijuana from this apartment. After entering, one of the officers observed, through the open doorway of the kitchen, a package of marijuana on a scale on top of the kitchen sink. The officers made the arrest, did a complete search of the apartment, and found additional marijuana. Justice Brennan indicated that the knock-and-announce rule is a Fourth Amendment requirement applicable to this case. Justice Clark, on the other side of the debate, reasoned that the unannounced entry to Ker's home fell within California's appended statutory requirement. Clark concluded that the officer's failure to knock and announce was not unreasonable under the standards of the Fourth Amendment. [28]

The knock-and-announce rule has already been given some leniency for exclusionary no-knock warrants. Recently the courts held (*State v. Johnson*, 2001) that to defend a no-knock warrant law enforcement is required to articulate a rational and specific suspicion that a no-knock entry is necessary to avoid destruction of evidence or to shield the officer serving the warrant from undue harm. The government needs to present the basis of that suspicion and frame the request on the totality of the circumstances. Although the server's opinion may be founded on familiarity and understanding, the government displays a minimal level of purpose to justify a no-knock entry warrant, which cannot be constructed based on a sheer hunch.

Sneak-and-Peek Warrants

A sneak-and-peek search warrant (also called a covert entry search warrant or a surreptitious entry search warrant) is a search warrant authorizing the law enforcement officers executing it to effect physical entry into private premises without the owner's or the occupant's permission or knowledge, and to clandestinely search the premises; usually, such entry requires a stealthy breaking and entering. [43]

In testimony to the Senate Select Intelligence Committee concerning the Patriot Act, Mr. Dempsey from the Center for Democracy and Technology, as a civil liberties representative, advocated that sneak-and-peek search provisions have been in existence for a number of years for intelligence investigations. He maintained that under the current Patriot Act legislation, the authority extended beyond terrorism has been abused, or at best, misapplied. The Patriot Act's provisions are being utilized in all federal criminal investigations. [35]

The Patriot Act has weakened court oversight of both search and seizure, including electronic surveillance. Critics say it has undercut existing notice protection to subjects of electronic surveillance, opening the door to potentially long-term secret wire-tapping. Former U.S. Representative Bob Barr, a Republican from Georgia who leads the Patriot Act reform organization, Patriots to Restore Checks and Balances, advances the concern of sneak-and-peek: "I think that the power that the government has under the Patriot Act ... is clearly contrary to the notion underlying the Fourth Amendment".[10] "The USA Patriot Act established a uniform nationwide standard for use of delayed-notice warrants to ensure an even-handed application of constitutional safeguards to all Americans".[14]

No Door, So No Place to Knock–Cyberspace?

Home Redefined

The physical home–a residence providing protection from the elements and segregating a perimeter between inside and outside–has traditionally held importance. The electronic influences on modern society have long generated fears regarding loss of privacy.[39] The correlation between an individual's house, private place, and personal spaces has become difficult to articulate and effectuate as Shapiro evidences:

> *The home is to some degree a private place and that a change in the border of home place/space also affects the boundary separating private from public. In other words, the boundary constituted by the home and the boundary between public and private are partially coincident. (p. 276)* [39]

Within the bounds of either a tangible house or a virtual space, an individual's private life is encapsulated with a legitimate expectation of privacy, and as such, is thought to be protected by the Fourth Amendment. In 1928, Justice Brandeis predicted:

Ways may someday be developed by which the Government, without removing papers from secret drawers, can reproduce them in court, and by which it will be enabled to expose to a jury the intimate occurrences of the home. ... Can it be that the Constitution affords no protection against such invasions of individual security? [53]

Technological developments may have turned Justice Brandeis' foresighted prediction into reality. A person's home is not only meant to provide a shield from nosy neighbors, but under the law is also designed to provide protection from a potentially intrusive government. Despite the healthy respect for the home privacy in American law, do those citizens with virtual homes in cyberspace have equal amounts of privacy protection from law enforcement?

"Cellphones and other portable electronic devices are, in effect, our new homes," the American Civil Liberties Union said in a court filing that urged the court to apply the same tough standards to cellphone searches that judges have historically applied to police intrusions into a home. [40]

During the second half of the 19th century, the boundaries of the home or residence have been strongly challenged. As telephone became commonplace, unsolicited calls intruded on an individual's or family's personal space. The popularity of the print media has forced journalism into spaces that were heretofore personal. The telephone, television, and now the Internet have become bidirectional flows of communication. The world outside the boundaries of the residence has entered the off-limit landscape and private matters that previously stayed behind closed doors have the potential of being broadcast across the world.[22]

The telephone was one of the first technologies that challenged the courts to establish a new paradigm for thinking about privacy issues. This shift was not rapid. *Olmstead v. United States* (1928) challenged the court. Previously, the court viewed the telephone as a device that communicated a person's voice outside the perimeter of the home and as such did not afford the same privacy protection under the law. The Olmstead decision was reversed by *Katz v. United States* (1967). The Katz decision, subsequently referred to as the Katz doctrine, changed the perception of inside as it pertained to the parameter of the home. Justice Harlan ruled that when Katz entered the public telephone booth and closed the door, his expectation of privacy was altered with this interpretation; the phone booth changed from a public space to a private place.

Cyberspace does not require a radical shift in how the police understand their function. While cyberspace may be an imperfect, inexact copy of the real world where the fit is imperfect, it remains analogous to nature (Huey, 2002). Therefore, if cyberspace is a parallel environment to the real world, it seems logical that individuals should be able to have a high expectation of privacy relative to their computers. This expectation may be heightened when those computers are located in their homes. A computer in a person's house is exactly the type of repository of personal information that the Fourth Amendment protects; just like it does a desk drawer. [53] Davis submitted that:

> *Most of us grew up in an environment where information was gleaned from the press, the education system, libraries and other places of learning–all physically tangible. But the Internet has meant that in a policing sense, evidence of crime has to be acquired from what people see or where information has been recorded in some way. (p. 51)* [56]

Hyde argued:

> *The Internet is a free and accessible entity where normal policing systems can be applied. ...The police' ability to respond to Internet crime is currently haphazard and based on luck rather than a prepared and researched provision of a service to the public. (p. 9)* [26]

Is There an Attack on Civil Liberties? A Paradox of Rights?

By the People

The Fourth Amendment establishes doctrine to protect citizens against unreasonable searches and seizures by government actors. These actors were typically law enforcement officers. The Fourth Amendment does not include protection against unreasonable searches by private parties. Recently a group of private citizens investigated suspected criminals by web searching on the

Internet. While these citizens were within their legal rights and were indeed helpful, their involvement raised Fourth Amendment concerns. *Burdeau v. McDowell* (1921) ruled that the Fourth Amendment is directed at state actions. Searches by private individuals are subject to the restrictions of the Fourth Amendment only if the private individual is acting as an instrument or agent of the government at the time of the search. In the case of *United States v. Reed* (1994), it was determined that if a private citizen is acting as an extension of, or an arm of the government, that citizen has crossed the agency line and is viewed under the same reasonable search and seizure interpretations of the Fourth Amendment as a police officer or any other law enforcement agency.

The case of *United States v. Miller* (1982) dealt with an individual as an *instrument or agent* as long as the government knew of and consented to the intrusive conduct, and the person performing the search intended to assist law enforcement efforts. There are others who have marshaled opposition against policing cyberspace. Groups such as the American Civil Liberties Union, the Electronic Frontier Foundation, and the Electronic Privacy Information Center have prepared for long drawn-out battles in court when they perceive that law enforcement has overstepped constitutional boundaries in enforcing crimes related to the Internet and cyberspace. [24] In addition, "Title II of the Electronic Communication Privacy Act–1986 (ECPA) prohibits private citizens from gaining unauthorized access to stored communication and enumerates specific procedural requirement for a government entity to gain access to store electronic communication".[53] The ECPA legislation is intended to establish protection for the intentional interceptions and endeavors to intercept, any wire, oral, or electronic communication. Therefore, the ECPA does not pertain to stand-alone computer systems because they are self-contained and do not transfer communications. Once connected to a network, the information communicated may fall under the language of the ECPA legislation.[17]

For the People

Recognizing that September 11, 2001, was a terrible event perpetrated on the United States, either to our soil or to our ideals, are we perpetuating the event by handing over our civil liberties to the government? Some would say yes. [31] Some would argue that those working to destroy our way of life have won and are continuing to gain ground on a daily basis without our conscious realization.

> *It's been true throughout American history that when the bullets fly, civil liberties are among the first casualties. During the Civil War, Abraham Lincoln suspended the right of habeas corpus, the constitutionally enshrined procedure by which a defendant can challenge a wrongful conviction. In World War II, Franklin Roosevelt interned 120,000 Japanese Americans and tried accused German saboteurs in military courts. The Bush Administration is leaning on these historical precedents in saying the traditional balance between security and freedom needs to shift, at least in the short term. We're an open society, President Bush declared last week, but we're at war.* [16]

Civil libertarians see rights being taken away by government in post September 11, 2001, in the name of a war on terrorism.[6] Many of the provisions of the Patriot Act can be used to carry out surveillance on ordinary citizens. The act gives the government expansive power to perform sneak-and-peek searches—to go into homes when the owner is not there and inspect the premise without informing anyone. Such searches are far more invasive than conventional knock-and-announce warrants permit.[16]

Fox asked, "Are we becoming a surveillance society? Sophisticated devices and techniques have greatly enhanced the capacity of government to intrude into lives of citizens."[21] It can be argued that in the United States we want to live in a society safe from child pornography, copyright theft, Internet scams, and terrorist plots.[20] If the outcome of safety and protection means a crackdown on cybercrime, some of the population believes that providing law enforcement additional latitude, even if it means a reduction in privacy rights, may be worth the sacrifice.[20] The opposing view[46] sees Internet surveillance by government, even for protection and safety, as an invasion of privacy. On the other side of the argument, if an individual is not doing anything wrong, why be concerned?

Against the People

Surveillance occurs long before individuals are arrested and able to defend themselves in a court of law with a jury of their peers. Elinson[18] reports that it is becoming common for police department officers to wear body cameras. Pundits will argue that this is to protect the American citizen from police abuse. Even the ACLU position themselves indicating cameras have the potential to be a protection of liberties for law enforcement as well as the private citizen when used with restrictions. At the same time, privacy concerns are raised.[18] American citizens have been protected by the constitution that provides for a fundamental belief of innocence until proven guilty. [32] The interception of electronic mail poses a wrenching concern. When and why should law enforcement be permitted to intercept a citizen's e-mail? By contrast, first class mail must contain evidence that it is involved in a crime before it is accessible. Having probable cause, a warrant can be issued. If an e-mail were to follow that same logic after delivery with probable cause existing, and only after a search warrant is obtained, should an officer be permitted to intercept the documentation?[9] Unfortunately, that logic does not exist in today's statutes.

E-mail surveillance can take place during the movement of digital traffic passing through the local Internet Service Provider (ISP) and possibly earlier. Tools that capture keystrokes as they are typed are used by criminals and peacekeepers alike.[13] The common citizen sees this technique used by companies that collect personal information. Individuals believe their anti-virus and other software exist to block the infringement activities. If a neighbor or a law enforcement officer used a telescope and watched someone write a letter, the person would have recourse under current laws. Should the same rights be provided for e-mail? Technology is in the hands of criminals and the government, while innocent citizens ride a rollercoaster filled with Fourth Amendment paradoxes. The future is apt to bring a new cadre of technology, with the need for legal interpretations relative to the impact these devices may have on the rights of all Americans, as well as the global population.

Theories, Laws, and Paradigms

Change Happens

Scientists examine a phenomenon according to a paradigm or worldview.[5] Paradigms gain their status because they are more successful than their competitors.[29] A changing paradigm does not mean the world itself has changed. Scientists may discover new and exciting instruments or see the world differently through new instruments and tools. [29] The decade of the 1990s launched a dramatic shift in the use of computers. Although the Internet was introduced before the graphic user interface (GUI) portal, now referred to as the browser; the paradigm shift to a GUI made the Internet assessable. In November of 1993, MOSAIC™, the first browser to include pictures, was marketed. NETSCAPE™ followed in 1994, and in 1995 Microsoft entered the scene with its initial version of Internet Explorer™. Access to the Internet created a drastic paradigm shift from the way banking, marketing, and advertising had been conducted. Along with the Internet's wide appeal and utility, privacy alerts such as SPAM and viruses are the byproducts of digitally encoded personal data; where cookies are left with every transaction. [57]

Electronic Communications Privacy Act

Defining cyberspace, networks, and freestanding computers, versus those that fall within the boundaries of the ECPA, leaves a great deal to interpretation and deciphering by the courts. The monitoring and interception of Bulletin Board Service (BBS) communications by law enforcement agents implicate two conflicting policy issues. On the one hand, the monitoring or seizing of communications by the government stifles the exchange of ideas. As Justice Douglas stated: "monitoring, if prevalent, certainly kills free discourse and spontaneous utterances"[50] "On the other hand, statutory and constitutional authority recognizes that law enforcement officials should be able to monitor communications that are otherwise freely accessible to the general public".[53] Search and seizure of digital evidence can be ambiguous, at best. The current laws regarding digital forensics remain in flux. Privacy rights as related to search and seizure of computer evidence generate additional complications. [24] One key concern is the concept of a deleted document. What an individual deletes from a computer is not necessarily gone, as it is in the physical world.

The Fourth Amendment and the ECPA find a legal line of delineation. The line is grounded in an individual's expectation of privacy. Public communications, those that are generally accessible by members of a broadcast group, have little expectation of privacy. Conversely, a message transmitted over a private network or transmitted to a private individual is accompanied by a high expectation of privacy, and these communications are therefore intended protection against unreasonable search and seizure by the Fourth Amendment. The ECPA sets out the provisions for access, use, disclosure, interception, and privacy protections of electronic communications. The law was enacted in 1986 and covers various forms of wire and electronic communications. This act defines electronic communications to include images, sounds, data, or intelligence of any nature transmitted by a domestic or foreign wire, radio, electromagnetic, photo electronic or fiber optic cable that affects commerce.[7]

Under the provisions of the Electronic Communications Privacy Act, an agent of the state can join a bulletin board service (BBS) and monitor messages, keeping his or her identity anonymous. A BBS user lacks the identity of other authorized users; therefore the agent does not raise any constitutional concerns. ECPA regulates unlawful access and disclosures of the contents that are governed by this law. The law also prohibits government agencies from demanding disclosure of these communications from an ISP without proper procedure.[7]

Precedence is growing as computer related case law involved with computer searches reaches beyond the trial courts and are decided in United States Courts of Appeal or at the State or Federal Supreme Courts level. Protection under the law develops as the courts and law enforcement recognize the differences between computers and other vaults of personal information. The courts need to understand the potential of searches and seizures of computers, as computer data encroaches into and potentially violates the aspects of an individual's professional and personal life.[53]

Patriot Act

The Patriot Act was legislated 45 days after the terrorist attack of September 11, 2001. The official title of the Patriot Act is the "Uniting and Strengthening America by

Providing Appropriate Tools Required to Intercept and Obstruct Terrorism (USA PATRIOT ACT) Act of 2001" (H.R. 3162). Even though much of the Patriot Act was a redefinition, expansion, or clarification of existing laws, the act greatly expanded several prevailing sections of law.

1. It increased the power of the FBI to compel people and businesses to turn over documents.
2. It expanded national security letter authority to allow the FBI to issue a letter compelling Internet service providers, financial institutions, and consumer credit reporting agencies to provide a record about people who use or benefit from their services.
3. It expanded another provision of law to authorize the FBI to more easily obtain a court order requiring a person to turn over documents or other evidence sought for an investigation, in order to protect against international terrorism or clandestine intelligence activities.
4. In both cases, the act removed the requirement that the above needed to pertain to an agent of a foreign power. This significantly expanded law enforcement access to records pertaining to Americans.[6]

With these expanded powers, the recipient of the subpoena is gagged forever from telling anyone, including an attorney, that they received the request. Mr. Dempsey, when testifying at the Senate Intelligence Committee, reminded the panel that broad authority can generate inappropriate investigations. Dempsey also noted that the Federal Bureau of Investigations (FBI) used that expanded power under the protection of the Patriot Act when they broke into a judge's chambers and did a sneak-and-peek search. Under the veil of the Patriot Act, the FBI also broke into a medical office to investigate a Medicare issue. These are nonviolent crimes, and yet government agencies are using the Patriot Act beyond its intended mission.[35] Privacy has also become more complex with the use of personal devices including mobile phones, performing work from home computers.

End Notes

1. Albrecht, K. (2002). Supermarket cards: The tip of the retail surveillance iceberg. *Denver University Law Review, 79*(4), 534-539 and 558-565.
2. Albrecht, K. (2005). *Wal-Mart hammered for controversial use of new technology: Dozens descend on Dallas store to protest RFID "Spychips."* Retrieved November 8, 2005, from http://www.spychips.com/protest/walmart/dallas-protest-press-release.html
3. Albrecht K., & McIntyre, L. (2004). RFID: Big brother bar code. *ALEC Policy Forum, 6*(3), 49-54.
4. Albrecht K., & McIntyre, L. (2005). *Spychips: How major corporations and government plan to track your every move with RFID.* Nashville, TN: Nelson Current.
5. Amaravadi, C. S. (2004). The laws of information systems. *Journal of Management Research, 4*(3), 130-136.
6. American Civil Liberties Union. (2003). *USA Patriot Act.* Retrieved April 10, 2006, from http://www.aclu.org/safefree/resources/17343res20031114.html
7. AOL, Inc. (2003). *Electronic communications privacy act.* Retrieved August 26, 2005, from http://legal.web.aol.com/resources/legislation/ecpa.html
8. AT&T. (2005b). 2002: *Privacy bird.* Retrieved October 28, 2005, from http://www.att.com/attlabs/reputation/timeline/02privacy.html
9. Bertron, D. (1971). Home is where your modem is: An appropriate application of search and seizure

law to electronic mail. *American Criminal Law Review, 34*, 163-195.

10. Bloomekatz, A. (2005, August 1). Drug-tunnel bust aided by controversial provision of USA Patriot Act. *The Seattle Times.* Retrieved August 12, 2005, from http://seattletimes. nwsource.com/html/localnews/2002413803_sneakpee k01m.html

11. Bovard, J. (1996). *Did the Supreme Court flush the Fourth?* Retrieved February 27, 2006, from http://www.fff.org/freedom/0296d.asp

12. Burdeau v. McDowell, 256 U.S. 465, 475, 65 L. Ed. 1048, 41 S. Ct. 574 (1921).

13. Chidi Jr., G. A. (2002). *Judge okays keystroke surveillance.* Retrieved April 10, 2006, from http://www.pcworld.com/news/article/0,aid,78284, 00.asp

14. Christensen, D. (2004). Justice watch: Courts. *Broward Daily Business Review, 46*(13), 1.

15. Chronology of data breaches reported since the choice point incident. (2006). Retrieved March 2, 2006, from http://www.privacyrights.org/ar/ChronDataBreach es.htm

16. Cohen, A., Dirkerson, J., Novak, V., Weisskopf, M., Fonda, D., & Winters, R. (2001). Rough justice. *Time Canada, 158*(24), 36-43.

17. Electronic Communications Privacy Act. (1986). Electronic Communications Privacy Act (ECPA) 18 USC §§ 2510−2520

18. Elinson, Z. (2014, August 16), More Officers Wearing Body Cameras. *The Wall Street Journal*, (A3).
19. Eddie, G. (2000). E-mail, the police, and the Canadian Charter of Rights and Freedoms: Retooling our understanding of a reasonable expectation of privacy in the cyber age. *International Review of Law, Computer, & Technology, 14*(1), 63-78.
20. Etzioni, A. (2002). Implications of select new technologies for individual rights and public safety. *Harvard Journal of Law & Technology, 15,* 257-290.
21. Fox, R. (2001). Someone to watch over us: Back to the panopticon. *Criminal Justice, 1,* 251-276.
22. Friedman, T. L. (2005). *The world is flat: A brief history of the twenty-first century.* New York: Farrar, Strauss, and Grioux.
23. Ghadar, F., & Spinder, H. (2005, July/August). IT: Ubiquitous force. *Industrial Management* [Electronic version]. Retrieved October 15, 2005, from http://www.allbusiness. com/periodicals/article/511919-1.html
24. Goodman, M. D. (1997). Why the police don't care about computer crime. *Harvard Journal of Law & Technology, 10,* 465-494.
25. Huey, L. J. (2002, July). Policing the abstract: Some observations on policing cyberspace. *Canadian Journal of Criminology,* 243-254.
26. Hyde, S. (1999). A few coppers change. *Journal of Information, Law, and Technology, 2,* 1-13.
27. Katz v. United States, 389 US 347 (1967).
28. Ker v. California, 374 U.S. 23; 83 S. Ct. 1623; 10 L. Ed. 2d 726; (1963).

29. Kuhn, T. S. (1996). *Structure of scientific revolutions.* Chicago: University of Chicago Press.
30. Lammers v. Nebraska. (2004). 267 Neb. At 688, 692, 676 N.W. 2nd at 727-28.
31. Lyon, D. (2001). *Surveillance after September 11.* Retrieved April 10, 2006, from http://www. fine.lett.hiroshima-u.ac.jp/lyon/lyon2.html
32. McCullagh, D. (2003). *Perspective: Guilty until proven innocent.* Retrieved April 10, 2006, from http://news.com.com/2010-1071-996625.html
33. Olmstead v. United States, 277 U.S. 438 (1928)
34. Park, M. (2005). *In Re charter communications: The newest chapter in P2P file sharing.* Retrieved March 5, 2006, from http://www.bu.edu/law/scitech/volume11 issue2/ParkUpdateWEB.pdf
35. *Patriot Act: Hearing before the U. S. Senate Select Intelligence Committee,* 107th Cong., 1 (2005) (testimony of Dempsey).
36. Perry, W. L., McInnis, B., Price, C. C., Smith, S. C., & Hollywood, J. S. (2013). *Predictive policing: The role of crime forecasting in law enforcement operations.* Santa Monica: Rand
37. Préfontaine, D. C. (2001, October). *Implementing international standards in search and seizure: Striking the balance between enforcing the law and respecting the rights of the individual.* Paper presented at the Sino Canadian International Conference on the Ratification and Implementation of Human Rights Covenants: Beijing, China.
38. Scheindlin, S. A., & Rabkin, J. (2000). *Electronic Discovery in Federal Civil Litigation: Is Rule 34 up to the Task?,* 41 B.C. L. REV. 327, 363.

39. Shapiro, S. (1998). Places and spaces: The historical interaction of technology, home, and privacy. *The Information Society, 14,* 275-284.
40. Sherman, M., (2014). Supreme court take on privacy in the digital age. Retrieved September 29, 2014 from http://news.yahoo.com/supreme-court-takes-privacy-digital-age-115436065--finance.html
41. Smith, R. M. (2001, February 19). Internet privacy: Who makes the rules? *Yale Journal of Law and Technology, 3.*
42. Smyth v. Pillsbury Co. 914 F. Supp. 97 (EDPa. 1996).
43. Sneak and peak search warrants and the USA Patriot Act. (2002, September). *The Georgia Defender,* p. 1
44. Stasser v. Yalamanchi, 669 So 2d 1142 (ct. Appl. La. 1996).
45. State v. Johnson, 168 N.J. 608, 619 (2001).
46. Steinhardt, B. (1999). *Law enforcement should support privacy laws for public video surveillance.* Retrieved April 10, 2006, from http://www.aclu.org/safefree/resources/16782res19990408.html
47. United States Constitution. Amendment IV. (1789, Ratified December 15, 1791). *Amendments to the Constitution of the United States.* Retrieved January 31, 2006, from http://www.gpoaccess.gov/constitution/pdf/con001.pdf
48. United States v. Miller, 688 F.2d 652, 658 (9th Cir. 1982).
49. United States v. Reed, 15 F.3d 928, 931 (9th Cir. 1994).

50. United States v. White, 401 US 745762-63 (1971) (Douglas, J., dissenting).

51. USConstitution.net. (2006). Retrieved January 31, 2006, from http://www.usconstitution.net/index.html

52. Weiser, M. (1991). The computer for the 21st century. *Scientific American, 265*(3), 94-101.

53. Winick, R. (1994). Searches and seizures of computers and computer data. *Harvard Journal of Law & Technology, 8*(1), 75-128.

54. Zubulake v. USB Warburg LLC, 55 FRS 3d 622 (SDNY May 13, 2003).

55. Mindrup, J.J. (2004). Knock-and-announce rule. *Creighton Law Review,* 38,97-153

56. Davis, C. W. (2004). Software for efficient file elimination in computer forensics investigations (Master Thesis, College of Engineering and Mineral Resources, 2004). (UMI No. 1423966).

57. McCullagh, D., & Mills, , M. (2006). Feds take fight to Google. *News.com* [Electronic version] Retrieved February 26, 2006 from http://news.com/feds+take+porn+fight+to+Google/2 100-1030_3-6028701.html

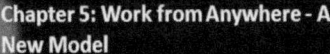

Chapter 5: Work from Anywhere - A New Model

Working anywhere introduces new challenges and business concerns framed within employee rights. The intersection of an employee's personal and work life and responsibilities has long been a difficult subject area for both employee and employer. Montana asserts "Blurring boundaries between what is work time and work equipment and what is personal time and personal equipment bring into question the ownership of information created on personal computers.

What expectation of privacy do employees have to personal information on employer-owned computers"? [14] Legal experts warn that employees should only have a "limited expectation of privacy" even if they send personal e-mails from private e-mail accounts at their workplace. As e-mail becomes a fundamental business tool in today's organizations, ethical and legal issues such as privacy rights and ownership will become paramount concerns. [8]

Background

Business articles, papers, and books exploring the issues of virtual companies, factories, and offices are increasing in frequency in both business press and in academic research.[1] However, very little if any work has been published on the questions of information ownership, privacy, and ethics and the impact to those constructs that occur when a worker transitions from an on premise employee to a Work-from-anywhere individual.

This chapter will evaluate the following questions as they relate to the remote worker-employer relationship and illustrate a social change:

- What rights do employers have to access employee-owned computers used for work purposes?
- What expectation of privacy do employees have to personal information on employer-owned computers?
- As e-mail becomes a legacy application in today's organizations, how are ethical and legal issues such as privacy rights and ownership reconciled?
- How does the Work-from-anywhere behavior model impact an employee's personal integrity and privacy rights?

The intersection of an employee's personal and work life and responsibilities has long been a difficult subject area for both employee and employer. Work is often done from home. The work may be performed on a computer supplied by the company or accomplished on the employee's own system. The question could be raised as to who owns the property rights to the work on the employee's personal computer.[13,14] If the work product was clearly created on the individual's own time, with his or her own computer, in his her own home the answer should be clear should it not? The employer may expect or assume that the organization has intellectual property rights to the contents that relate to the work tasks, at least to some degree. The employee may have thoughts on the other side of the paradoxical quandary.

Intellectual Property rights and employee privacy are only two of many complex doctrines entwined with the Work-from-anywhere model that are being challenged; liability concerns, tax implications, and residency considerations are other implications. This review will not ponder tax and residency issues. It will also only briefly touch on liability. Although working mobility has a tremendous incentive it comes with challenges to both the employer and the employee. [15] It should also be noted that what is legal to either party may not be what is ethical. "Because telecommuting challenges traditional notions of how work is structured, it poses unique questions in the area of employment law."[23] In the process of answering the above questions, this chapter will provide the reader an integration of data on remote workers and related employment doctrines.

The Melding of Private and Work Life

The boundaries between the workplace and everywhere else have become blurred in the recent years.[5] Broadband capabilities to the home, wireless networks within the house, the car, a coffee shop, and software such as VPN that enable users to get access to their company's computer from anywhere all contribute to a transition and integration of office activities into personal lives. As computer networks incorporate fiber optics, cable modems, DSL, and satellite into a digital matrix, cyberspace will offer more and more possibilities for remote working.[1] Cellphones have G4 capabilities and autos have Wi-Fi as optional equipment. The office is extended outside of the controlled office environment. Current technology voids many of needed mental relaxation.

Friedman asserts that society is undergoing a process of perfecting two new dimensions that computers have added to communication: the visual and the interactive. Dwellings have become more than a place to spend time after study or work. Homes, cars and other wireless "hotspots" are now the school, social gathering place, and the workplace.[5] Although working outside of the office has tremendous advantages, it comes with its challenges as well. One of the biggest problems with working from home is not being able to switch *off* when you walk out of an office at the end of the day.[15]

What is *new* in the evolving organizational landscape, however, is the increased assimilation of so much of an employee's private life activity into systems that are supplied on the desktop. While the telephone, along with company e-mail, were systems that an employee may have used in the past to conduct personal business, the employee now has access to the Internet, cell phones, PDAs and laptops, and connectivity from virtually anywhere. This cadre introduces a remarkable new range of potential personal uses that may legitimately interfere with productivity, or even expose the employer to legal action and financial liability.[24,26]

Working-from-anywhere introduces additional business concerns including the use of personal computers on the company network and the use of company owned equipment for personal use. Some businesses are negligent about the use of employee-owned computers on their networks. GreenLine Systems, Inc is an example. GreenLine supplies risk management software for the transportation supply chain. The company has subcontracts with the Transportation Security Administration to ensure the safety of cargo entering U.S. seaports. GreenLine should be the poster child for an ironclad security policy. The company does sell its expertise to its clients. But the vice president of risk management and GreenLine's general counsel acknowledges that virtually everyone at the company uses personally owned laptops that may or may not be secure. [6]

Use Policies

Acceptable use policies are developed and maintained to articulate what the company considers proper use of the Internet, e-mail and general computer usage. The policy should describe the employee's privacy expectations on private use and personal e-mail. If the organization has remote users, the policy should also be clear about the use of home computers on the network and the implication of personal use and the ownership of intellectual property for works created on the home computer. A written policy will also allow the company to spell out alternative resolution procedures in case of a dispute.[11] Raysman and Brown advance the policy discussion by when they argue:

> *Formal e-mail policies to serve as a guideline regarding the acceptable use of company e-mail, and today, 79 percent of all U.S. companies have such written policies. Typically, these guidelines both remind employees that e-mail should be used primarily for business purposes and set out disciplinary procedures for abuse of this policy. Due to an increase in regulatory and legal scrutiny of e-mail, a majority of employers now use some form of discipline to enforce these policies.[19]*

In designing e-mail and Internet use policies, it's important to clearly educate employees that review/monitoring will occur and that they should have no expectation of privacy when using the company's computers and networks. A policy should not brandish *Zero tolerance.* Employees should know that occasional personal use is fine, but content must not violate equal opportunity or anti-harassment policies. Brandon writes, the e-Policy Institute says some employers ban all personal use of company e-mail systems, while others allow unrestricted personal use. Firms will allow personal use but within limits, e.g. 30 minutes, during a workday or only during breaks or before or after normal business hours.[2] Mack asserts "taking steps to reduce the likelihood of technology related disputes is certainly an important component of any risk management program".[11]

Employer Obligations

Often when the issue of privacy is discussed, the need for employee rights is the sole focus. Privacy laws and employment regulations have traditionally been strange bedfellows. They frequently cross paths in their purpose and execution.[12] With many recent laws enacted, the employer too is under obligation to adhere to regulations that protect the privacy of clients and even step further down the information supply chain - the clients of clients. As an example, the Health Insurance Portability and Accountability Act referred to as HIPAA obligates anyone processing patient data to make sure that anyone who shouldn't see hospital data doesn't. If a patient billing company has Work-from-anywhere employees, this obligation extends well beyond the wall of the hospital, doctor's office, or even the billing company. Financial institutions have specific requirements under the Gramm-Leach-Bliley Act of 1999 that became effective in 2001. This legislation requires financial organizations to inform customers of their privacy policies. At the state level, California has enacted legislation that requires companies to alert their customers if data is suspected of being compromised.[6] Often employers are seen as the opposition when it comes to monitoring and other privacy issues. Employees are frequently unaware of the organization's legal requirements.

Why Employers Monitor

The courts generally hold employers responsible for providing a workplace that is free of harassment, sexual harassment, bullying, gossip, or reputation defacement. To reduce the risk of litigation employers are increasingly monitoring their employees' use of cellphones, voice over IP (VOIP), PDA, laptops, and e-mail.[20] Human resource literature of the 1990s was replete with alerts that employers implement technology to monitor employee work performance. Employers utilizing software programs track network usage, read e-mail, and smart badges that may even pinpoint physical location. These methods and actions can extend to the homes of work anywhere employees.[16]

The Electronic Communications Privacy Act of 1986 is the basis for employers' use of e-mail monitoring. Although the legislation prohibits intentionally intercepting an e-communication or accessing a stored message it has a number of exception for employers. For example it does not apply to the interception of e-mail in route or between the sender and the recipient. The exception does also the review/monitoring of e-mail saved on a user's server or hard drive.[21]

"The growth of an entire industry dedicated to Internet tracking has evolved out of the legitimate concern for employee on-the-job behavior. Employees have stolen proprietary information while planning to establish competing business. Employees have downloaded sexually explicitly material for their own enjoyment or to harass coworkers, creating potential liability for the company itself". [3] Generally speaking, companies are not legally compelled to inform their employees that their e-mail and other transactions is being monitored. "Although e-mail systems in the workplace are common, the laws addressing employee rights and employer monitoring rights are ambiguous".[4] To combat the abuse of e-mail, an estimated 60 percent of companies in the United States, have adopted some type of computer software to monitor their employees' electronic communications. Employees whose e-mails have been monitored sometimes seek court intervention. To date, the legal remedies available to the employee appear to be limited, at least with respect to e-mails sent over a work e-mail system. Employees also have attempted to use state privacy laws to prevent employers from accessing their e-mails, but have not had any real success with this strategy. In one case, *Smyth v. Pillsbury Co.* 1996, a federal district court held that the employer was free to read the employee's e-mail messages without giving any warning to the employee.[19]

Another state court ruled in *McLaren v. Microsoft Corp.*, 1999, that an employee did not have a reasonable expectation of privacy in the contents of e-mail messages that the employee stored in a *personal* folder on his computer that required a password to open. The court found that the employer provided the computer to the employee so that he could perform this job, and therefore the employee did not have a reasonable expectation of privacy in his work e-mail messages.[21] To date, the orthodoxy in state privacy law is that there is no expectation of privacy for e-mails sent at work through an employer's e-mail system.

Often is it not the company that initiates the request for data from employee records. More and more, the courts because of an action that the company is involved in, order the business to turn over instant messages, e-mails and computerized memos as evidence. Anything is fair game in electronic discovery -- even the records stored on an employee's home computer, his PDA, his cell phone, his laptop, his iPod. It's not unusual for a company of any size to hear: Produce every e-mail sent between one date and another for a specific person, e.g. Mary Jones. If the party served with the discovery order can't retrieve the messages it could be subject to hefty fines or worse -- allegations of destruction of evidence. In some cases, those allegations have led juries to set punitive damages in the millions of dollars. [22]

Expectation of Privacy

Some believe personal privacy is under siege, with employers perusing e-mail their employees send from the personal computer or personal assistant. [10] Legal experts warn that employees should only have a "limited expectation of privacy" even if they send personal e-mails from private e-mail accounts at their workplace. Employee privacy expectations can be defeated and/or eliminated by employer policies. Currently, employees understand from employer policies that their use of other company equipment such as telephones and computer systems can be monitored at any time by their employer and that the employer has access to their telephone calls, Internet sites visited, and e-mail messages when they use that company equipment. Company-provided cell phones and PDAs are no different. Increasingly, privacy relates to the diverse modes by which people, personal information, certain personal property, and personal decision-making are made less accessible to others. Though privacy is protected by law, it is also governed by culture, ethics, and business and professional practices. Given the multifarious nature of privacy, constitutional principles and legislation are only the first step toward understanding the effects of increased technology in our lives and the potential harms and benefits it carries.

The classic legal argument for the harm affected by the loss of privacy was stated by Samuel D. Warren and Louis D. Brandeis in 1890: "The intensity and complexity of life, attendant upon advancing civilization, have rendered necessary some retreat from the world, and man, under the refining influence of culture, has become more sensitive to publicity, so that solitude and privacy have become more essential to the individual; but modern enterprise and invention have, through invasions upon his privacy, subjected him to mental pain and distress, far greater than could be inflicted by mere bodily injury".[25] Nelson, advanced the theory that the Right to privacy is grounded in pre-constitutional concepts and possesses both legal and moral underpinnings. In part, Brandeis and Warren developed their argument from the natural law concepts that dominated discussions of law in the late 1800s. The creation of legal doctrine was to mirror the dictates of natural law. In line with this logic of the relationship between mankind and governance, Brandeis and Warren constructed their argument around a notion of privacy that was founded on human dignity and equal respect for persons. In their articulation of the "inviolate personality," they sketch out a notion of privacy that is still relevant today. Privacy is an essential component of

113

personhood and is limited only by sufficient justification for the invasion of privacy by government, private individuals, or business entities. While Brandeis and Warren provided a framework that came to characterize the development of privacy doctrine, any questions regarding the practical implementation of privacy remain problematic throughout the development of the doctrine of privacy.[25]

"There can be no expectation of privacy when using e-mail or instant messaging at the office -- not even with Yahoo ™ or other public accounts," said Ted Demopoulos, a consultant and founder of Demopoulos Associates in Durham, N.H. "Anything that goes over the network can and may be monitored -- it's that simple".[10]

Ownership: The Intellectual Property Question Incorporated in the Copyright Act of 1976; ownership initially resides with the author or authors of the work. In the broad interpretation, it is the person or persons that actually created the work or translated the idea into a fixed and tangible expression that is due the copyright protection. The act did include however provisions for "works for hire" which considers the employer or the entity for whom the work was prepared to be the author who owns the copyright.[7] Work-for-hire implies that the rights to the works were never owned by the creator.[9] To determine whether a work is for hire under the Act, it should be ascertained whether the work was prepared by an employee or an independent contractor. In determining whether a hired party is an employee under the general common law of agency, consideration needs to focus on the hiring party's ability to control the manner and means by which the work product is produced. The "work for hire" doctrine and the study of the principles of agency are beyond the scope of this discussion. Note must be given however, the exception rule for academics. The general rule for determining when an employee's act is within the scope of employment – when it is of the nature the employee was hired to perform. There is great debate for example over the teacher exception. Few reported rulings exist that take on the issue of whether the institution or faculty member own copyright to teaching materials prepared by a faculty member. Cases have tended to follow the "teacher exception" to the work for hire doctrine.[7]

Chapter Wrap-up

Work Product Model

The literature has supported the conclusion that several legal doctrines create decisional boundaries around the research questions. These boundaries interact and counteract with each other. A summation of the literature can be viewed in a "Right v. Work Product" model (Figure 12) depicting the three domains/lens creating output: "work for hire", Employee, and Employer. These three domains generate output for and become the organization's intellectual property. The intellectual property carries rights that the three domains relinquish for payment of work performed.

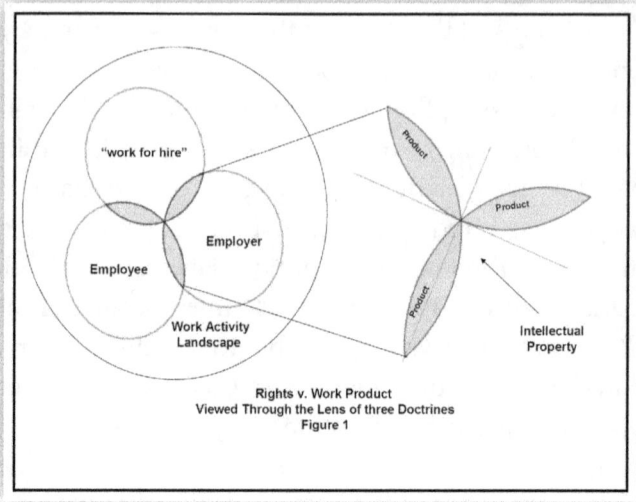

Rights v. Work Product
Viewed Through the Lens of three Doctrines
Figure 1

Figure 12: Rights v Work Product ©2005 Ponschock

The model also exhibits the relationship of the three rights doctrines to each other and the workspace in which they transact. Despite some case law on the subject of e-mail or other transactional movement in the workplace, there is still uncertainty and ambiguity in the law about the extent to which employers may monitor employee e-mail or other digital activity. In such circumstances as transmitting instant messenger messages at work, or for "outsourced" e-mail systems, or for e-mails sent over a personal account such as Yahoo! or Hotmail through the company's Internet connection, or using a home computer for business purposes. As there still may be potential liability for employers, employers must remain careful about monitoring e-mail, and employees should be equally cautious about e-mails they send at work.

Expectation of Privacy

Although employees assert they have a privacy right in e-mail sent and received on company equipment, the courts have concluded to the contrary. There is no reasonable expectation of privacy in e-mail sent, stored or received at work (Smyth v. The Pillsbury Co., 1996).[28] The computer hardware and software belong to the employer, and so does all the information stored on it. The Smyth court noted that the e-mail communications were made voluntarily over the company e -mail system and that the company was not requiring the employee to disclose any personal information about himself. Not-withstanding any assurances by the employer that such communications would not be intercepted, no reasonable expectation of privacy could be found. Moreover, the court held that the company's interest in preventing inappropriate and unprofessional comments or even illegal activity over its e-mail system out-weighs any privacy interest the employee may have had in comments made over e-mail.[19]

Employee privacy must not be taken lightly. The organization must act in the interest of all employees. In that landscape the organization both for ethical and legal motives must maintain safe, and harassment-free environments. To that end, with or without the employee's knowledge, the employer may need to reviews/monitor e-mail or other transactions that pass through or is stored on company equipment or equipment used by an individual while an agent/employee of the organization. These transactions or work product may be created as the work performance of the employee or private output. This is unknown until reviewed or monitored and may be a transaction that will put the employer and/or other employees at risk.

If the person initiating the transaction was a contractor hired for a specific purpose to create a specific product, again the employer has the responsibility to ensure the quality of the transaction and that the transaction is within the contractual parameters of the "work for hire" agreement. If the contractor used the environment of the organization to produce the work product that contractor should also have no expectations of privacy. If the work product falls outside the work for hire agreement for any reason, e.g. not what was contracted, not what the agreement specified, or not within the time parameters there is a decision point on whether that work resides in the intellectual property domain of the employer.

Ownership Matrix

All transactions that fall outside the work product landscape are not easily deciphered and usually fall into courts to clarify the paradox or dispute. Three constructs supported by literature and previous research that can push a transaction outside of the work activity landscape in Figure 12 they are time, location, and scope. In the time construct, the product may have been created on the individual's own time and may then be the individual's work product and as such must be treated differently by the employer. This product may have been produced on the employer's equipment but outside the normal work time parameter which would have different results than an "out of scope" product produced during work hours.

118

Figure 13 is a matrix that assists in understanding this inconsistency.

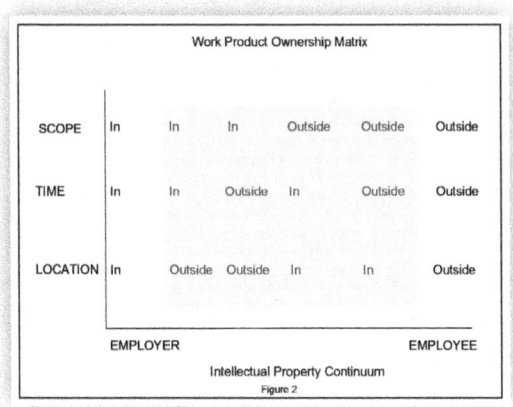

Figure 13: Intellectual Property Continuum
© 2005 Richard Ponschock

Those cases that fall within the employment scope of work, during work hours and on company premises or locations specified and agreed upon between the employee / contractor and the employer are generally easy to resolve. Transactions or products produced outside the scope of work that the individual was hired to produce, not within work hours and not on company property are not as straightforward. Company property may be defined as location equipment or network, and is again easy to argue. Literature supports the argument that those transactions in the grey area have no case law, arbitration results, or clear legal foundation and must be resolved on individual merits.

What rights do employers have to access employee-owned computers used for work purposes? Literature and the

119

courts will generally support the monitoring and possibly the discovery of documents that were created in the course of business. Using the previous discussion it is recommended that the employee keep all personal transactions on a separate hard drive to prevent confusion should review, monitoring or discovery become necessary.

What expectation of privacy do employees have to personal information on employer-owned computers? It appears very clear that employee should harbor no expectations of privacy on transactions created and/or stored on an employer's equipment or transacted through the organizations network.

As e-mail becomes a legacy application in today's organizations, how are ethical and legal issues such as privacy rights and ownership reconciled? Although employers are under no legal obligation to inform employees that their e-mail or other transactions can be reviewed or monitored, the establishment of an acceptable use policy that clearly articulated the guidelines of what is permitted and what is not can prevent conflict or resolve conflict if it occurs. Employees need to understand that the employer is under the obligation to create a safe and harassment-free work place for all employees and adhere to government regulations.

How does the Work-from-anywhere behavior model impact an employee's personal integrity and privacy rights? The acceptable use policy can help establish parameters, but the burden falls more heavily on the employee to function without the immediate direction of a supervisor.

Going Forward

There are many aspects of the "Work-from-anywhere" employee that remain open for future research. Some businesses show 4 to 25 percent increase in productivity.[27] The question must be asked; what about the other side of the scale; does the procrastinator become less productive? Do some Work-from-anywhere employees become forgotten; even overlooked for other positions because the employees on premise are more visible? Employers are required by law to provide a workplace that is free from recognized hazards that are causing or likely to cause death or serious physical harm".[17] What if an employee is injured at home, is it a workman's comp claim. Although working from home is not new to the business scene, many questions remain under-investigated as the cyber era unfolds.

Security issues are also serious concerns whenever and wherever computers are used organizations must to additional measures with remote users.[18] Work anywhere employees extend the security and privacy concerns of the employer. What if the employee house is burglarized, what if the communications are not secured, and what if an employee crosses to the dark-side of the ethics. As society transitions along the digital continuum, it is essential to explore how humanity itself is evolving – transforming.

End Notes

1. Barnatt, C. (1995). Office space, cyberspace & virtual organization. *Journal of General Management, 20*(4) p78-91.
2. Brandon, G., (2005, February 21). Workers' web use: A growing concern for more employers. *Kipling Business Forecasts*, 2005.
3. Bupp, N. (2001). Big brother and big boss are watching you working. *USA, 5*(2), 69-81.
4. Dombrow, J.C. (1998). Electronic communications and the law: Help or hindrance to telecommuting? *Federal Communications Law Journal, 50*(3), 685-709.
5. Friednam, A. (2004, July 25). Internet has changed the way we use our homes: the line between home and workplace is expected to become more blurred. *The Gazette (Montreal, Quebec)*, A11
6. Gardner, E. (2003, November). To catch a thief. *Corporate Counsel*, 10, 108.
7. Holmes, G., & Levin, D.A. (2000). Who owns course materials prepared by a teacher or professor? The application of copyright law to teaching materials in the Internet age. *Brigham Young University Education & Law Journal*, 1, p165-190.
8. Holmner, M., & Britz, J.J., (2001). Electronic mail ethics: A technological perspective. *South African Journal of Library & Information Science. 67*(2)
9. Joss, M.W. (1995). Copyright in the cyber age. *Computer Artist, 4*(5), 27.
10. Koprowski, G.J. (2005, February 23). The web: Online privacy under attack. *UPI Science News*
11. Mack, L.R. (2000, Fall). Have I told you lately that I LOVE YOU? A proactive approach to the misuse of

computer technology resources in the business environment. *Journal of Alternative Dispute Resolution in Employment*, 2(3),11-14.

12. Mignin, R.J., Lazar, B.A., & Freidman J. M. (2002). Privacy issues in the workplace: A post-September 11 perspective. *Employee Relations Law Journal, 28*(1).

13. Montana, J. C. (2005 May/June). Who owns business data on personally owned computers. *The Information Management Journal, 39*(3), 36-42.

14. Montana, J.C., (2005). Who owns Business Data on Personally Owned Computers? *Information Management Journal, 39*(3), 36-42.

15. Montero, P., (2004 November). Two perspectives on how to work from anywhere. *The Journal for Quality and Participation, 27*(3).

16. Nolan, D. R. (2003). Privacy and profitability in the technological workplace. *Journal of Labor and Research, 24*(2).

17. OSHA (1999). The occupational safety & health act of 1970, 29 U.S.C. Para 654(a)(1)(1999), 29 C.F.R. Para 1903.1

18. Phelan, S. (2004). Home is where the office is. *Journal of Accountancy, 194*(6).

19. Raysman, R., & Brown, P. (2005, April 12). Computer law: Workplace e-mail and the electronic communications privacy act. *New York Law Journal*, p3.

20. Shumaker, T. A. (2003, November). Employee privacy versus employer rights. *Nursing Homes Long Term Care Management, 52* (11), 60-63.

21. Sidbury, B.F. (2001, July). You've got mail... and your boss knows it: Rethinking the scope of

electronic communications privacy act. *Journal of Internet Law*, 5(1), 16-23.

22. Smith, E. D. (2005, June 6). Virtual evidence. *Akron Beacon Journal*, 029950606

23. Swink, D. R. (2001). Telecommuter law: A new frontier in legal liability. American Business Law Journal, 38, 858-900.

24. Townsend, A.M., Alberts, R.J., & Whitman M. E. (2000, March) Employer liability under the communications decency act: Developing effective policy response. *Employee Responsibilities and Rights Journal*, 39-46.

25. Warren, S. D., & Brandies, L. D. (1890). The right of privacy. *Harvard Law Review*, 4(5).

26. Whitman, M. E., Townsend, A.M., & Alberts, R.J. (1999, June). The communications decency act is not as dead as you think. *Communications of the ACM*, 42, 15-17.

27. McCune, J. C., (1998). Telecommuting revisited. *Management Review*, 87: 10-16.

28. Smyth v. Pillsbury Co. 914 F. Supp. 97 (EDPa. 1996).

Chapter 6: My Digital Avatar

Mankind is in a transformational era. This paradigmatic shift is rooted in the underlying adoption and use of rapidly evolving information technology with attendant sociological behaviors. Based on this more dramatic shift, it is posited that it is more than just a transformational process; rather a speciation event that is underway dictated by the technological metamorphosis and the avant-garde adoption from a sociological perspective. It is rapidly becoming apparent that additional psychological and sociological issues may become evident in such research, but the focus here remains the identification of the technological identities that have, and are evolving, and their contribution, termed "The Avatar Effect" to the posited speciation event. Additional research may be warranted from a psychological and sociological perspective.

Background

It can be hypothesized that humankind and its physical society is transforming into a digitally grounded species living in two worlds – planet earth and cyberspace. C. Otto Scharmer advanced the theory that "transformation must pull us into an emerging possibility and allows us to operate from that altered state rather than simply reflecting on and reacting to past experiences."[21] There is a contention here that the current state in the digitization of our society is beyond a point of retraction. Kuhn referred to a transformation of this magnitude as a "Paradigm shift".[12] This chapter expands the concept introduced in a previous introductory work that if society continues on the continuum of the Information Communication Technology (ICT) dynasty it will reach beyond just a transformational event to a more dramatic societal speciation event.[16]

Identity

Identity has been defined and explained numerous times in various ways. Dictionaries vary with some of their definitions. Combining the number of defining attempts, for this paper, identity is who you are, the attributes that define you to yourself and others – your essential self. It can be further argued that identity is contextual based. We can take on different personae depending on the environment at a point in time. Following this theory, the ability to have multiple identities is possible. In the digital world, personae attached to actors in specific cyber settings are referred to as "avatars". "The fundamental theme of avatars pertains to the strain of research on

126

digital identity construction via avatar creation in technology mediated environments."[10] Jin has empirically studied the impacts of avatar-to-avatar communication as contrasted to face-to-face communication. Jin posits that "When people migrate from the real world to an avatar-based virtual world, they may perceive a discrepancy between the actual self and the virtual self."[10] This is important in the context of this research as it contributes to the growing theory of the paradigmatic sociological impact, and ultimately contributory towards the speciation event and the ultimate digitization of society.

The sociologist Manuel Castells of the University of California Berkeley in his book "End of Millennium", argued that our current societal environment was formed in the late 1960s and early 1970s: from among other drivers, the information technology revolution and a new culture of "real virtuality" and this environment can be referred to as the digitization of society.[5]

What do Alice in Wonderland, Pinocchio, and Power Rangers have in common – they transform into another "persona" - they change identity. Although these are fictional, humankind is also grounded in identity. Identity can be described as sameness between two potentials. Society in general has evolved around identity. Phrases like:

- Having an identity crisis
- Identity theft
- Identify yourself
- Provide two forms of identity
- Change my identity

- I wish to keep my identity private

are commonly used euphemisms. The term identity is derived from the French word 'identité' and originated from the Latin noun identitas, -tatis. The Latin noun matured out of the adjective idem meaning "the same". This derivation emphasizes the sharing of a percentage of sameness or oneness. Identify is best constructed as being both relational and contextual, while the act of identification is best viewed as an inherently processing activity.

As the digitization of society transforms humankind into a digital "person", keeping an identity private; is difficult at best. Empirical research has found that "after years of social networking, cellphone usage, and numerous other digital generators, the humankind will face an environment that has a biography of their personal data collected and aggregated." [15] In that same study it was reported that "illusions of privacy or the belief that we have a right to privacy will diminish" (p.109). Scott McNealy, the former CEO of Sun Microsystems stated "there is no privacy; get over it."

To many, "getting over it" has given way to attaining multiple personas through the channels of social media. These new identities may be as simple as a creative blog identifier (i.e. Flash80) or the persona of a player with a graphical representation on Internet worlds (i.e. Second Life or Farmville). This avatar effect has been empirically studied by Sherry Turkle and published in the text *Life on the screen: Identity in the age of the Internet*. She states:

A rapidly expanding system of networks, collectively known as the Internet links millions of people in new spaces that are changing the way we think, the nature of our sexuality, the form of our communities, our very identities.[20]

Who am I (The digital me)

Virtual Reality is the impact of digital revolution. While making it possible to create different dimensions of the reality and providing more experience to the self, there is a kind of a danger in virtual reality. Twenty-first century society is part of a collectivity of social networks to which we tie various self-created attributes developing contextual personas. The ties to these "digital self's" may be weak or strong.[7,8] Social media established ties may be with actors who do not even exist.

Identity is both a means of control and a means of self-definition. We can shape, reinvent create multiple identities.[22] We can be "Alice" on both sides of the looking glass. In the virtual worlds of cyberspace multiple digital identities are common. "In the real-time communities of cyberspace, we are dwellers on a threshold between the real and virtual, unsure of our footing and inventing ourselves as we go along." [20] Technology is an identity "creator". Technology has the power to make us – and to break us and increasing numbers "live" in the digital world. Jin stated "the divergence between the actual self and the virtual created personae point to the belief that avatar users perceive the virtual self to be more physically attractive than the actual self. This finding implies that

129

avatar users project the ideal self (the virtual ideal self) by creating an avatar more physically attractive than the actual self."[10]

As one thinks about "who they are", the line between the real person (or the "offline" one) and the projection onto social networking sites (the 'online" self) is becoming blurred. There are situations where real person has multiple unique personalities in the virtual world.[9] These unique and ever-growing personalities continue to contribute to the sociological paradigmatic shift underway, ultimately becoming a major contributor to the posited speciation event.

Social Networking (beyond virtual)

Berger and Luckman advanced the theory of social construction when they noted that everyday life is "not only taken for granted as reality by the ordinary members of society in the subjectivity meaningful conduct of their lives, it is a world that originates in their thoughts and actions and is maintained as real."[3] A shift is occurring in the way members of our technologically oriented society interrelate.[14] It is changing from one of personal relationships between individuals to virtual chats. Technology has been driven by the "new Media" over the past 10 years (Figure 14).[15] A distinguishing attribute of a true technological reconstruction is that many innovations occur at about the same time.[11]

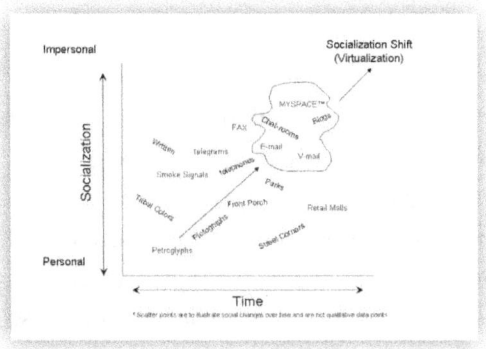

Figure 14: Socialization Paradigm Shift © 2007 Ponschock

Social networking enabled by Internet and gains in Information Communication and Technology - ICT is performing crucial roles throughout society. It is the enabler of societal digitization. The Internet and cloud architecture has become the encyclopedia, telephone, photo album, recipe box, and much more for more than 2.3 billion users worldwide. It is also the stage on which humankind can take on numerous personas. A recent Nielsen report indicates "80 percent of active Internet users spend nearly a quarter of their total time at social networking and blog sites."[6]

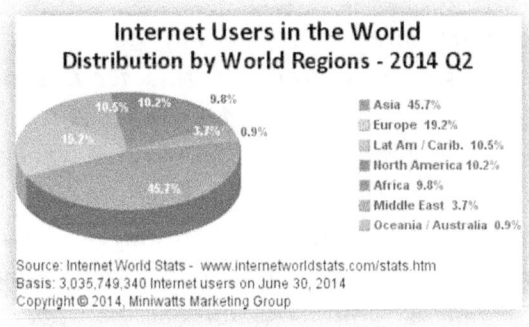

Figure 15: Global Internet Use

131

While the electronic form of social interaction has its limitations, the convenience and ease-of-use of social networks through blogs, forums, and e-mail-lists can lead to a society that is actually much more connected than ever before. Social media has now allowed this sphere of friends and colleagues to become global. These social networks will only serve to enhance relationships, and to benefit society as a whole. In doing so, these users are making use of "cloud computing," an emerging architecture by which data and applications reside in cyberspace, allowing users to access them through any web-connected device .[9] The age of the global community is upon us and enabled through advancements with technologies such as ICT. Consider as evidence the global classroom consisting of individuals around the world cooperating towards their desired objectives in real time!

Advances in brain research and multisensory perception could play an important role in the development of virtual relationships. Neural devices already allow people to control electronic equipment such as wheelchairs, televisions, and video games via brain–computer interfaces. One day soon, avatars may also be controllable this way. Virtual reality may become so advanced that it could trick the brain into thinking the invented images it is responding to are real—and human emotions would follow accordingly. Avatars will cause people to feel love, hate, jealousy, etc. And as haptic technologies improve, our abilities to respond physically to our virtual partners will also improve: Sexual pleasure may be routinely available without any inter-human stimulation at all. If it becomes possible to connect virtual reality programs directly to the brain, thoughts and motions may also be digitized, rendered binary and reduced to 0s and 1s. Feelings of satisfaction and pleasure (two key components in any relationship) could be created between avatars without any "real" stimulus at all. But would they be real or mimetic? Once humans begin to perceive virtual social interactions as actually having occurred, it will greatly impact individuals, relationships, communities, and society as a whole.[4]

Digitization of society is rapidly progressing beyond chatting, e-mail and games. Web sites such as Ancestry.com, and DNATribes.com offer inexpensive DNA tests. It is reported that this ongoing research explores the genetic links between cultures around the world, challenging old ideas about ancestry and identity. DNA, can combine genealogy, medical technology, and social networking. The future of falling in love may also be online. Dating websites are no longer "taboo". These social networking communities are an accepted way to meet and connect with new people, through the convergence of information technologies and digital entertainment. By utilizing and enacting the avatar in all of us, it remains a conundrum as to whether you are truly connecting with the real you!

In the new Virtual Society, we will see an increasing transition from basic matchmaking sites to sites that enable people to actually participate in online "dates" without ever leaving their armchairs.

The Digital You

Identity can be compromised in many ways through a collection of digital traits, facts and biometrics. Fingerprints, DNA, and facial recognition make up the "digital you". In October, 2012, 4,200 students at a School in San Antonio were given identification cards with radio frequency identification (RFID) chips. The card transmits GPS like location information via the RFID microchip to electronic readers throughout the campuses. The school district argues that "This is non-threatening technology... this is not surveillance".[13] Ronald Stephens, executive director of the nonprofit National School Safety Center, said he didn't believe the technology to be widespread but predicted "it'll be the next wave" in schools. The chips use radio-frequency identification (RFID) transmitters and only work on campus. Although the school administration does like to call the chips "tracking" — a clerk in the office can find out if a student is on campus, and count him/her in attendance.[1]

The FBI's Next Generation Identification (NGI) program will include biometrics such as iris scans, DNA analysis and voice identification in its arsenal in addition to established facial recognition.[19] Tests in 2010 showed that the best algorithms can pick someone out in a pool of 1.6 million mug shots 92 per cent of the time.[19] Government is not the only societal entity experimenting with facial recognition. Social media is also digitizing our persona. Facebook has rolled out a facial recognition system globally, to the protests of privacy regulators. The feature identifies 'people' in photos as they are uploaded and suggest to the user uploading the photo that they 'tag' the friends identified by the facial recognition system.[17] A person's **face** is the one link between offline and online identities, argued Acquisti, associate professor of information technology and public policy at the Heinz College and a Carnegie Mellon CyLab researcher. "When we share tagged photos of ourselves online, it becomes possible for others to link our face to our names in situations where we would normally expect anonymity." [23] Pattern recognition software can identify individuals using unique biometric characteristics such as face recognition, iris, scan, hand geometry and the possibility of DNA scanning.[18]

In today's educational environment, online programs are continually experimenting with ways to ensure that students are the ones doing the work in these virtual settings. Many of the aforementioned technologies are under significant scrutiny as to their ability to potentially enable more stringent identification of these virtual students. This is a key technological shift to the ever-increasing issues surrounding who the person may actually be at the other end of these virtual engagements.

Chapter Wrap-up

This chapter triangulates three views providing a foundational grounding toward the discussions on the actors' multiple digital personas as society follows the continuum toward the "Digitization of Society: The Avatar Effect". Conducting daily transactional activities in the physical world generate transactions that are collected, stored in digital vaults, and mined for personal and commercial purposes.[15]

Digitization affects that have already influenced the way society and it's actors function. Turkle posits that identity, the "multiple self" is ensconced between the eroding boundaries of real and virtual environments. She states "from scientists trying to create artificial life to children *morphing* through a series of virtual personae, we shall see evidence of the fundamental shifts in the way we create and experience human identity."[20] Jin's "Virtual identity discrepancy model" (VIDM) empirically examined the social psychological mechanism underpinning avatar-based virtual self-representation and virtual communication.[10] Beniger, a seminal thinker on society and evolutionary control posited

> *As in earlier revolutions in matter and energy technologies, the nineteenth-century revolution in information technology was predicated on, if not directly caused by, social changes associated with earlier innovations; because technology defines the limits on what a society can do, technological innovation might be expected to be a major impetus to social change* [3]

The next chapter will analyze how reputations are impacted along the transformational journey.

End Notes

1. Associated Press, (2012 November 27). Track suit: Family challenges 'locator' chips embedded in student ID cards at Texas schools. Retrieved January 10, 2013, from http://articles.washingtonpost.com/2012-11-27/national/35509762_1_student-id-badges-school-bathroom-texas-schools

2. Beniger, J.(1986). The control revolution: Technological and economic origins of the information society. Cambridge MA: Harvard University Press.

3. Berger, P.L., & Luckman, T. (1967). The social construction of reality: A treatise in the sociology of knowledge. Garden City, NY: Doubleday.

4. Brown, A., (2011). Relationships, Community, and identity in the new virtual Society. *Futurist. 45*(2), 29-34.

5. Castells, M., (2010). *End of millennium: The Information Age: Economy, society, and culture* Volume III. Oxford, UK: Blackwell Publishers.

6. Dortch, M., (2012). Dortch on SaaS & cloud computing: The blog. Retrieved May 21, 2012, from http://dortchonsaas.blogspot.com/2011/11/inventory-management-and-mobile-social.html

7. Granovetter, M.S. (2004). The impact of social structures on economic development. *Journal of Economic Perspective, 19* (1), 33-50.

8. Granovetter, M.S. (1973). The strength of weak ties. *American Journal of Sociology, 78*(6), 1360-80.

9. Hongladarom, S., (2011). Personal identity and the self in online and offline world. *Minds and Machines (21)*, 533-548.

10. Jin, S.A. (2012). The virtual malleable self and the virtual identity discrepancy model: Investigative frameworks for virtual possible selves and others in avatar-based identity construction and social interaction. *Computers in Human Behavior, 28*(6), 2160–2168.

11. Kranzberg, M. (1989). IT as revolution: The information age. In T. Forester (Ed.). *Computers in the human context* (pp. 19-32). Cambridge, MA: MIT Press.

12. Kuhn, T.S. (1996). *Structure of scientific revolutions.* Chicago: University of Chicago Press.

13. Miller, J. R., (2012). Student-tracking system at Texas schools prompts privacy concerns. Retrieved February 12, 2013, fromhttp://www.foxnews.com/us/2012/09/12/texas-school-district-defends-use-student-tracking-mart-id-card/#ixzz2ETorDCuy
http://www.foxnews.com/us/2012/09/12/texas-school-district-defends-use-student-tracking-mart-id-card/#ixzz2EToAZkIo

14. Mumford, L. (1970). *Myth of the machine: The pentagon of power.* New York: Harcourt.

15. Ponschock, R. L. (2007). Computer technology, digital transactions, and legal discovery: A phenomenological study of possible paradoxes (Doctoral dissertation, Capella University, 2007) (UMI No. 3246872).

16. Ponschock, R.L. & Becker, G.F., (2011). CLOUD Technology: A transformational dynasty on the ICT

evolutionary continuum and contemporaneously a societal speciation event. *European Journal of Management.* ISBN: 1555-4015.

17. Purewal, S., (2011). Why Facebook's facial recognition is creepy. Retrieved March 7, 2013 from http://www.pcworld.com/article/229742/Why_Face books_Facial_Recognition_is_Creepy.html

18. Rowley, W. (2002). Surveillance society and transparent society: *New challenges of society. Journal of State Government,* 2(2), 16.

19. Readon, S., (2012) If the feds fit the face. *New Scientist,* 215 (2880), 20.

20. Turkle, S. (1995). Life on the Screen: Identity in the Age of the Internet. New York: Simon & Schuster.

21. Scharmer, C. O. (2007). *Theory U: Leading from the future as it emerges.* Cambridge. MA: Society for Organizational Learning.

22. Shroff, M. & Fordham, A. (2010). Do you know who I am?: *Exploring identity and privacy. Information Policy,* 15, 299-307.

23. Acquisti, A., (2011). More than facial recognition. Retrieved May 1, 2015 from http://www.cmu.edu/homepage/society/2011/sum mer/facial-recognition.shtml

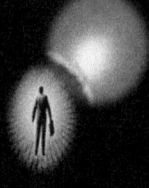

Chapter 7: Reputation in the Digital Era

Reputation as explored and examined herein, has evolved to a state of continual rapidity requiring an exhaustive and comprehensive understanding and attention to keep pace with potential pitfalls and exposures. Historic mitigation of barriers to technological innovation has been reinforced through this research initiative, and has taken on greater impact to the evolving digital personal profiles that are emerging for individuals and businesses.

Cottage industries related to data mining, reputation management, and improved personal/technological security considerations have evolved and are growing quickly. Follow on research related to digital "crumbs" and their requirement for proactive attention and management has evolved and is an increased area for exploration and examination.

Background

Cassio in William Shakespeare's play Othello intones:

> *Reputation, reputation, reputation! Oh, I have lost my reputation! I have lost the immortal part of myself, and what remains is bestial.*[9]

Reputation describes one's honor, integrity, and dignity knitted into a singular vale. Solove submits "We currently live in a world where extensive dossiers exist about each of us. These dossiers are in digital format, stored in massive computer databases by a host of government agencies and private-sector companies."[46] The aggregate of the dossiers formulate the essence of our reputation. Svantesson reported "Protecting one's reputation has arguably become harder in this time of *youtube*, 'blogs', and mobile phone cameras."[48] Dictionaries vary some in their definition of reputation. Combining the number of defining attempts for this paper, reputation is the overall quality or character as seen or by the public in general. This definition is not new to the digital society. What is new is the social media that is storing and conveying the characteristics that establish the reputation of individuals and organizations. Reputations built over decades can be changed or destroyed in a virtual moment. The following are several examples illustrating the reputation impact of digital media:

E-mail is at the center of the scandal that brought down CIA Director David Petraeus, one of the country's most decorated generals.[15]

"I am here today to again apologize for the personal mistakes I have made and the embarrassment I have caused," Weiner further summarized that the announcement brought swift relief to his Democratic colleagues, who had become increasingly uneasy as details emerged about Weiner's online contacts with women — including his sending of explicit photos of himself to them over Facebook and Twitter.[17]

Superintendent Nancy Sebring's earlier-than-expected departure from the Des Moines public schools was prompted by the discovery that she had carried on sexually explicit conversations using her district e-mail account, some of them during the workday, school officials confirmed Friday.[47] She had violated the district's technology policy, which prohibits use of school computers and e-mail accounts for personal use or the exchange of sexually explicit materials.[7]

A decorated Army captain has been relieved of command after his exchange of sexually explicit e-mails with former Des Moines schools superintendent Nancy Sebring.[23]

This chapter is a continuation related to the paradigmatic shift[24] based on technological and societal transformations associated with "digitization".[35] As the transformation progresses, reputation becomes a primary construct in the "digitization" of society in general and mankind specifically.

"Digitization" has distorted the boundaries of reputation

conveyance. Prior to the digital era an individual's reputation may have been communicated through friends, cohorts, clients, or individual actions. "Gossip is no longer the resource of the idle and vicious, but has become a trade, which is pursued as an industry as well as effronter."[51] Justice Brandeis was reacting to the new technology of the time, the mass-circulation newspaper. [13] The Internet has accelerated the ability to spread gossip well beyond that of a mass-circulation newspaper. Technology and the Internet have introduced communities that do not exist in geography and have no tangible physical presence. These virtual villages or townships [36] are not represented by geography, social class, or financial accounting. Instead, their cyber position is defined and driven by curiosity.[26] A century ago, social interactions involved relationships with others who were within a short walking radius.[13] Companies advertised on radio or local papers, many by word of mouth. Deals were struck with a handshake in the local coffee shop. Organizational size permitted employees to more readily see the whole. For many, especially in the industrialized West, small face-to-face communities are disappearing.[35] Reputations can and are being created, and altered by strangers, individuals you have never met. "Details about our private life on the Internet can become permanent digital baggage."[45] Gossip can tarnish an individual's or company's reputation; it exists as a bundle of half-truths and incomplete tales. While campfire stories dwindle as the fire goes out; Internet ashes are left behind forever.

As the "digitization" of society transforms humankind into a *DIGIPERSON*, maintaining an accurate reputation has become the origin for a new social discipline –

reputation management. "Our reputation matters quite a lot to us, but it also matters a lot to others in society, who use it to determine whether to trust me."[45] Character references are melting away as a requirement of the past. Social media including Facebook, and YouTube (accurate or not) are the "documents" of record. To the requestor this "is" your reputation.[45] Individual and company reputations are interwoven in the digital web of contextual data. The "digital person" can be, has been or can impact the trust shield protecting reputation.

Reputation can be contextually defined[39] and socially impacted. Social context theory is based on the interplay between social forces that affect individual behavior and individual and group actions that change society. While reputation is personal, it also has a socially phenomenological grounding. Berger and Luckman advanced the concept of social construction when they noted that everyday life is "not only taken for granted as reality by the ordinary members of society in the subjectivity meaningful conduct of their lives, it is a world that originates in their thoughts and actions and is maintained as real."[4] A paradigmatic shift [24] is occurring in the way members of the digitally transformed society interrelate.[30] It is changing from one of personal relationships between individuals to virtual chats. Ponschock argued that technology has been driven by the "new Media" over the past 10 years (Figure 1) [33, 21]

Social networking enabled by the Internet and advances in Information Communication and Technology - ICT are performing transformational roles across all segments of society.[35] These constructs are enablers of societal "digitization". Social media is an environment in which humankind can display, create, or tarnish reputations. While the electronic form of social interaction has its limitations - the convenience and ease-of-use of social networks through blogs, forums, and e-mail-lists can lead to a society that is actually much more connected than ever before. Social media has now allowed this sphere of friends, colleagues, and enemies to become global. These social networks can enhance or destroy reputations.

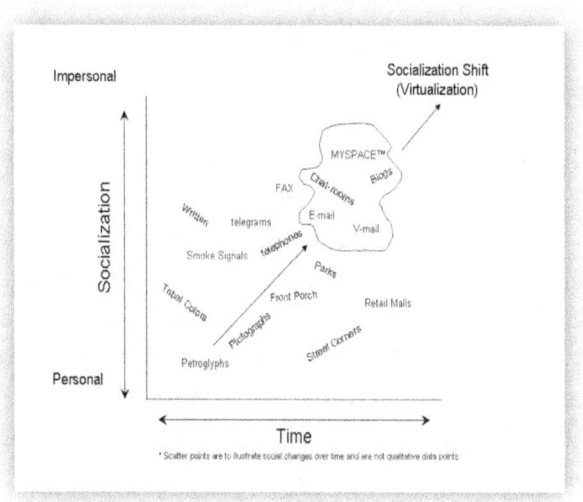

Figure 16: Socialization Paradigmatic Shift
© 2007 Ponschock[33]

Business Reputations

Gibson, Gonzales and Castanon theorized that "Reputation is arguably the single most valued organisational asset."[16] "Corporate reputation is increasingly regarded as a highly valued, intangible asset that is difficult to imitate and accordingly may provide a sustainable competitive advantage."[37] Approximately 65 percent of major corporations do not monitor social networking web sites like Facebook, YouTube, or the video-oriented MetaCafe.[20] A Ponemon Institute survey of 2011 posited that 85 percent of the respondents indicated that one of the five most important factors impacting an organization's reputation was Internet and social media communications. According to the survey results, reputation is only two categories lower than a company's concern with its financial health. Ninety two percent of these same respondents indicated that data privacy and security was extremely important to the security of the company's reputation.[32]

Even the best customer service will not guarantee that satisfaction of every customer.[40] As an example; a law firm in Florida was shocked when it found that everyone searching for legal services and the firm of Rumberger, Kirk & Caldwell also displayed Lying Scumbags, a web-site by Billy R. Kidwell, a former plaintiff who sued General Motors, concerning an alleged defective pickup truck. The derogatory link was visible for everyone to see. Anyone checking for the law firm online couldn't miss the ugly slur.[41] The law firm did not have the ability to execute a defense or have its 'day in court'.

Companies conducting e-commerce [6] must be prepared for the possibility that one day they will encounter the painful realization that when "Googled" your company is portrayed less than factual because of a derogatory post that appears on search engine results from Google™ , Yahoo, Bing or extraneous blogs. In fact, hundreds of thousands of dollars of sales are lost each day as a result of false, erroneous or misleading search engine results. Whether the negative content is from a competitor, a news site, or a blog, the impact can be financially troubling. Reputation Management Consultants can resolve many issues but the financial losses that may never be known combined with the cost of consultants can be permanently lost. "While the Internet can be a powerful tool in enhancing a company's reputation, it also is fertile ground for information that can damage, or, in some cases even destroy a company's **reputation**." [1]

"Social media has evolved into a radicalized moral compass, capable of destroying a Brand in seconds." [52] A positive online reputation is critical in the field of digital commerce. Perspective customers and future business partners will conduct research before sealing a deal.

Individual Reputations – How Can You Manage What You Can't Control

"What is honor?" asks Falstaff in *Henry IV*, Part 1. "Honour is a mere scutcheon ..." That shield can be easily tarnished or removed totally in the digital cyber-sphere. Your online reputation may reflect who you are or it may paint a picture of someone that you would not recognize. Remember, your digital anthology becomes your online reputation. It has been reported that 24 percent of college admission advisors have checked social media content while reviewing a candidate. More startling, 70 % of company recruiters have rejected potential employees basing on online findings. "Your public self is exactly that. It is what is seen in the virtual public eye and your public self reflects not only the actual you, but the perceived public image of you." [19]

A 2005 story now almost a legend started when a woman was on the subway in her native South Korea when her dog decided that this was a good place to do its business. The woman made no move to clean up the mess, and several fellow travelers got agitated. Yes, she should have been criticized and humiliated on the spot. However, the devastation of "Poop Girl's" reputation did not stop when the subway stopped. A passenger took a picture of her with a cell phone camera, it posted on a popular Internet website and from that point it went viral. "Within days, her identity and her past were revealed. Requests for information about her parents and relatives started popping up and people started to recognize her by the dog and the bag she was carrying."[22] Shortly after the cyber-sphere of blogs, taunts, and inquiries the "Poop Girl" even quit her position at a University. Accurate or inaccurate rumors can spin out of control.[46]

In the Internet /cyber world a "Catfish" is someone online who's really pretending to be someone else for the objective of fostering a romantic relationship. This could be using social media sites such as Facebook and Twitter, as well as on Internet dating sites. A recent ABC News Headline read "Notre Dame: Football Star Manti Te'o Was 'Catfished' in Girlfriend Hoax." Te'os reputation was shattered. We will never know if this was the reason the Notre Dame football star finished second for the Heisman Trophy or did not get picked in the first round of the NFL draft.[10] What is known is that Te'o and his tarnished reputation will live forever on the cybersphere of gossip. Te'o was reported as saying "This is incredibly embarrassing to talk about, but over an extended period of time, I developed an emotional relationship with a woman I met online." Te'o continued "We maintained what I thought to be an authentic relationship by communicating frequently online and on the phone, and I grew to care deeply about her. ... To realize that I was the victim of what was apparently someone's sick joke and constant lies was, and is, painful and humiliating." Notre Dame's athletic director, Jack Swarbrick reportedly stated: "But the thing I am most sad of is—" he added, pausing to apologize and wipe away tears, is "that the single most trusting human being I have ever met will never be able to trust in the same way ever again."[10] He fell for a girl he never met based on an invented social media presence.

"Catfish" is an example of how the "digitization" of humanity is rapidly progressing beyond chatting and e-mail. Reputations can be affected by what is mined from web sites beyond social networking. Personal mining can take place in advanced site such as Ancestry.com. DNATribes.com offers inexpensive DNA tests that may prove a different lineage. It is also reported that this ongoing research explores the genetic links between cultures around the world, challenging old ideas about ancestry and identity. DNA can combine genealogy, medical technology, and social networking.

Digital Tolerance

A 2010 PEW research report found that online reputation-monitoring via search engines increased from 47% in 2006 to 57 %. This 10% increase would appear to conflict with the findings that 33% of Internet users say they worry about how much information is available about them online, down from 40% in December 2006. Concern is down but monitoring is up.[27] Worry over personal information varies considerably among age groups (Figure 17).

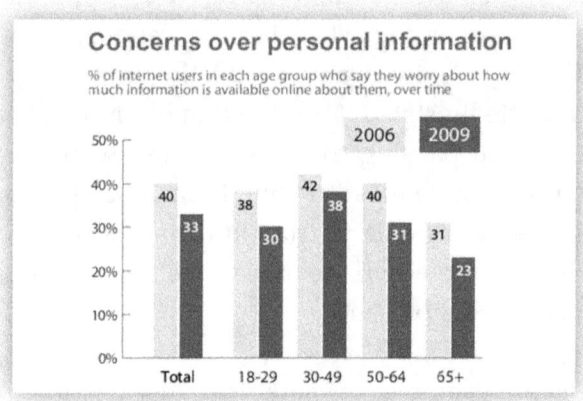

Figure 17: Personal Information Concerns[27]

The 2009 research show concerns over personal information are down across all groups. Tolerance for digitization is producing immunity across all ages. This tolerance can be contrasted with a 2011 Ponemon Institute study showing that 92 % of the corporate management respondents indicated that reputation or brand protection was important or very important.[32] These divergent illustrations would indicate that we are more tolerant of our personal reputation than of our corporate image (Figure 18).

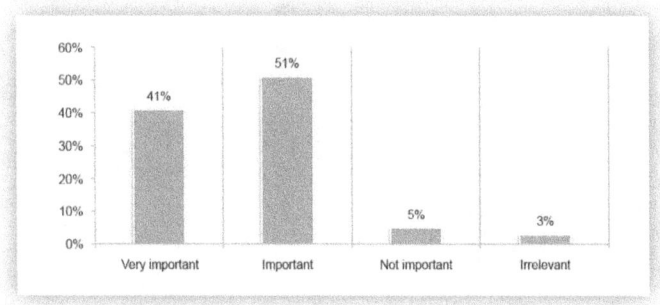

Figure 18: Reputation vs. Image[32]

Reputation Management

Struggles to manage reputations are becoming a 21st century skill and emergent industry. Reputation management is also referred to as Online Reputation Management or the acronym ORM. ORM involves monitoring online information characterizing you or your company and eradicating the damage if possible. [31] Reputation systems attempt to "unsqueeze the bitter lemon".[38]

Reputation systems are both content-driven and user-driven.[11] Both categories of reputation systems rely on feedback theory (Figure 19).

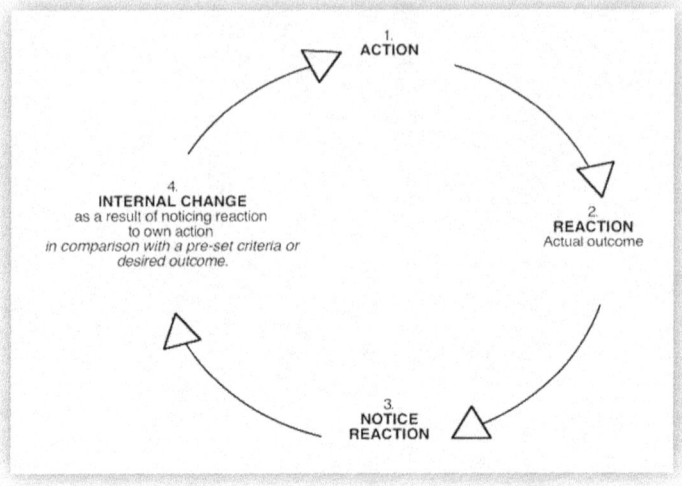

Figure 19: Simple Feedback Loop[11]

From an individual's viewpoint, a feedback loop exists when my system notices how the world responds to my behaviour and I adjust my behaviour in response that response, and so on. At any point my options are to do more or less of what I am already doing, or change to a different kind of behaviour. Either way, my aim will be keep things as they are or to change things to get more of what I want and less of what I don't want. However, you cannot get a systemic perspective from an individual's point of view. In a healthy system I take into account that other people are doing the same, that we are all part of a wider system that no one part can control. It is interesting to note that the use of the word 'feedback' in everyday language has been accompanied by a shift of attention.[49]

Content-driven management is based on the analysis of transactions and input responses. User drive systems can be related to the feedback sought on sites like Amazon where the buyer's rating of the transaction or service is solicited. This type of feedback can be biased.[11]

Figure 20: Amazon Feedback Form

"Reputation systems seek to establish the shadow of the future to each transaction by creating an expectation that other people will look back on it. The connections among such people may be significantly weaker than in transactions on a town's Main Street, but their numbers are vast in comparison." [38]

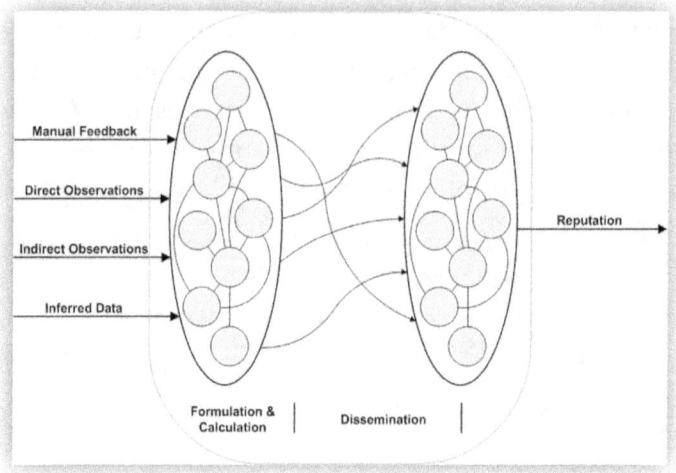

Figure 21: Depiction of How a Reputation System Works[18]

Figure 21 portrays the general model of a reputation system. The model shows the input to the system comes from multiple sources. Based on the source, the system produces a reputation metric through the use of an algorithm. Once calculated, reputation metric values are then disseminated. The large ovals encapsulate the reputation system itself, performing as interpreter and the distributed system.[18]

Reputation management takes on two distinctive features; observation and extermination. It is challenging at best and may be impossible in the worst-case scenario to eliminate inaccuracies or even find the initiator of the fallacy. Individuals and businesses can be defined by what appears on Google, Yahoo and Bing. "In fact, hundreds of thousands of dollars are lost each day because of false,

154

erroneous or misleading search engine results. Whether the negative listings are from a competitor, a news site, or a message board, the impact can be financially challenging at best and devastating at worst."[42]

Reputation systems are intended to contribute in the protection of the buyer. A good seller reputation has been shown to reduce transaction-specific risks and therefore generate price premiums for sellers. [2] Family, friends, and neighbors are traditionally asked first for reputation feedback before hiring services like contractors, plumbers, or gardners.to find out if they can recommend someone for the job. Over 1.5 million households use the reputation feedback system "Angie's List". This system follows a subscription model. With a pay-to-use model, filtering is more likely. Feedback opinions will come from clients who actually use the services of the supplier unlike Craigslist or the yellow pages. Figure 22 illustrates a sample of an "Angie's List" feedback form.

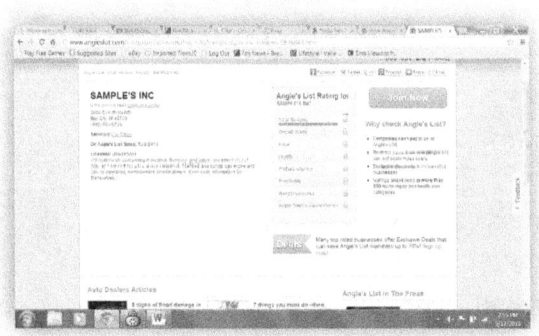

Figure 22: Angie's List Feedback Form

Chapter Wrap-up

Reputation can be compromised in many ways through a collection of digital traits, facts and biometrics. Fingerprints, DNA, and facial recognition and digital content make up the "digital you". Some of the digital characteristics fabricating an online reputation were examined in a 2010 PEW research report.[27] Figure 23 illustrates what we think others can see.

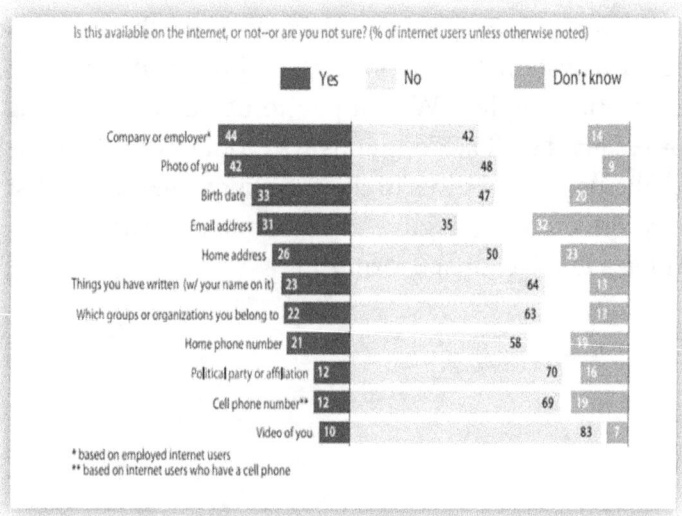

Figure 23: What We Think Others Can See[27]

More recently, reputation systems have been proposed as a means to filter out ersatz content (pollution), a means to identify and provide incentives for quality submission and a way to create résistance to others.[11]

This meta-analysis triangulates multiple studies providing a synthesis of numerous dialogues on the impact of digital reputations as society. "Digitization" has influenced the way society and it's actors function. Beniger, a seminal thinker on society and evolutionary control theorizes "As in earlier revolutions in matter and energy technologies, the nineteenth-century revolution in information technology was predicated on, if not directly caused by, social changes associated with earlier innovations" (p.9); "Because technology defines the limits on what a society can do, technological innovation might be expected to be a major impetus to social change".[3]

Having synthesized these studies, it is the authors' contention that the real-world and cyber-worlds has become interwoven establishing the foundation of the "digitization" of society. Internet reputation is more important than ever. Schiller reports

> *According to a Microsoft survey of more than 1,200 hiring managers in December 2009, 79% of companies and recruiters reviewed online information about job applicants and 70% had actually rejected candidates based on what they found.* [44]

> *Understanding how online reputations form, evolve, and dissipate is far from simple. And in this era of ever-present hyper focused social networking, people frequently spread information about themselves with little thought or*

restraint, only to watch it spread across
the Internet in days or even hours. [44]

Reputation is only one of numerous data points that contribute to the eventual transformation[43] of society and humankind. This report is the continuum of the journey society and humankind is traveling toward the speciation event that will transform the technological milieu to the digital being. Several data points have been presented. Figure 24 is a diagrammatic representation of transformational convergence. Ongoing correlational research connecting the remaining data points in the "digitization" phenomenon is necessary.

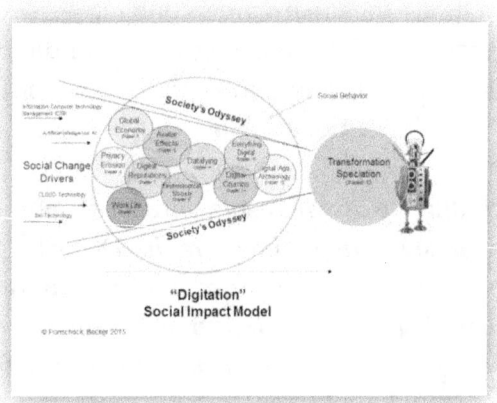

Figure 24: Digitization Convergence Model

"Digital "crumbs are the epicenter on the transformation continuum. It was discovered during this and previous studies that the trail of "digital" crumbs (i.e. identity, privacy, and utmost personal data) are pervasive and warrant an examination and report separate from the data points they affect.

158

End Notes

1. Akin, C. J., & Pinker, L.T. (2010, February). *Protecting your company's reputation on the Internet.* Paper presented at the meeting of ABA Section of Litigation Corporate Counsel CLE Seminar.
2. Ba, S., & Pavlou, P.A. (2002). Trust can mitigate information asymmetry by reducing transaction-specific risks. *MIS Quarterly Volume, 26*(3).
3. Beniger, J. (1986). *The control revolution: Technological and economic origins of the information society.* Cambridge, MA: Harvard University Press.
4. Berger, P.L., & Luckman, T. (1967). The social construction of reality: A treatise in the sociology of Knowledge.Garden City, NY: Doubleday.
5.
6. Berman. S.J., & Bell. R. (2011). *Digital transformation: Creating new business models where digital meets physical.* Somers, New York: IBM Institute for Business Value.
7. Bhasin, K. (2012, June 5). *Be careful with your work e-mail, so you don't end up getting burned like this former superintendent.* Retrieved June 20, 2013, from http://www.businessinsider.com/case-study-former-superintendent-nancy-sebrings-e-mails-2012-6#
8. Briggs, J. (2012). *Social context theory.* Retrieved May 15, 2013, from http://www.ehow.com/about_5414476_social-context-theory.html#ixzz2PiNNGk24
9. Crowther, J. (2005). *No fear Othello.* Retrieved September 13, 2013, from http://nfs.sparknotes.com/othello/

10. Curry, C., James, M.S., & Harris, D. (2013 Jan 16). *Notre Dame: Football star Manti Te'o was 'Catfished' in girlfriend hoax.* Retrieved June 16, 2013, from http://abcnews.go.com/US/notre-dame-football-star-manti-teo-dead-girlfriend/story?id=18232374

11. de Alfaro, L. A., Kulshreshtha, A., Adler, B., & Pye, I. (2011, August). Reputation systems for open collaboration. *Communications of the ACM, 54*(8), 81-87.

12. Dortch, M., (2012). Dortch on SaaS & cloud computing: The blog. Retrieved May 21, 2012, from http://dortchonsaas.blogspot.com/2011/11/inventory-management-and-mobile-social.html

13. Ermann, M. D., Williams, M. B., & Shauf, M. S. (1997). *Computers, ethics and society.* New York: Oxford University Press.

14. Evolution (2002). The theory of evolution – part II. Retrieved May 1, 2012, from http://h2g2.com/dna/h2g2/alabaster/A737985

15. Gaudin, S., (2012). *E-mail lessons from Gen. Petraeus' downfall: It may be easier than you think to trace e-mails, so be mindful of what you are sending.* Retrieved September 19, 2013 from http://www.cso.com.au/article/442164/e-mail_lessons_from_gen_petraeus_downfall

16. Gibson, D., Gonzales, J.L. & Castanon, J. (2006). The importance of reputation and the role of public relations. *Public Relations Quarterly, 51*(3), 15-18.

17. Hernandez, R., (2011). *Weiner resigns in chaotic final scene.* Retrieved June 1, 2013 from http://www.nytimes.com/2011/06/17/nyregion/anth

ony-d-weiner-tells-friends-he-will-
resign.html/?pagewanted=all&_r=0

18. Hoffman, K., Zage, D., & Nita-Rotaru, C. (2009). A survey of attack and defense techniques for reputation systems. *ACM Computing Surveys, 42*(1), 31.

19. Ivester, M., (2012). *lol... OMG! What every Student needs to know about online reputation management, digital citizenship and cyberbullying*. Reno, NV: Serra Knight.

20. Jackson, R. A. (2009). Keeping your reputation clean: Corporate missteps and evolving risks can leave organizations open to negative stakeholder perception and destroyed trust. *Internal Auditor. Institute of Internal Auditors, Inc.* Retrieved October 25, 2013 from HighBeam Research: Http://www.highbeam.com/doc/1G1-201869432.html

21. Kranzberg, M. (1989). IT as revolution: The information age. In T. Forester (Ed.). *Computers in the human context* (pp. 19-32). Cambridge, MA: MIT Press.

22. Krim, J., (2005. July 7). Subway fracas escalates into test of the Internet's power to shame. *The Washington Post.*

23. Krogstad, J.M., (2012, July 7). Updated: Army captain relieved of command over Nancy Sebring e-mails. Retrieved September 16, 2013 from http://www.desmoinesregister.com/article/2012070 7/NEWS/120707004/Updated-Army-captain-relieved-command-over-Nancy-Sebring-e-mails

24. Kuhn, T.S. (1996). *Structure of scientific revolutions.* Chicago: University of Chicago Press.

25. Levin, A., & Abril, P. (2009). Two notions of privacy online. *Vanderbilt Journal Of Entertainment & Technology Law, 11*(4), 1001-1051

26. Luthra, N. (2006). *The "Real" and the "Virtual" in public space.* (Master Thesis, University of New York at Buffalo) (UMI 1431955).

27. Madden, M., & Smith, A., (2010). *Reputation management and social media: How people monitor their identity and search for others online.* Retrieved May 20, 2013 from http://pewInternet.org/Reports/2010/Reputation-Management.aspx

28. Miller, J.R., (2012). *Student-tracking system at Texas schools prompts privacy concerns.* Retrieved September 12, 2013 from www.foxnews.com/us/2012/09/12/texas-school-district-defends-use-student-tracking-mart-id-card/#ixzz2ETorDCuy

29. Miller, J. R. (2013). *Big Brother: LA police sued over massive data collection gleaned from cameras.* Retrieved September 12, 2013 from http://www.foxnews.com/us/2013/05/10/la-police-sued-over-massive-data-collection-gleaned-from-cameras/#ixzz2Su5MCjvZ

30. Mumford, L. (1970). *Myth of the machine: The pentagon of power.* New York: Harcourt.

31. O'Reilly, T. (2013). *Introduction to online reputation management for financial services professionals.* Amazon, Kindle edition

32. Ponemon (2011). *Reputation impact of a Data breach: U.S. study of executives and managers.* Ponemon Institute LLC

33. Ponschock, R. L. (2007). *Computer technology, digital transactions, and legal discovery: A phenomenological study of possible paradoxes.* (Doctoral dissertation, Capella University, 2007) (UMI No. 3246872).

34. Ponschock, R.L,. & Becker, G.F., (2010). Virtuality of boundaries: - the iceberg in the current business curriculum: A mandate for systemic thinking. *Association for Global Business Proceedings, 22,* Paper 29. ISSN: 1050-6292.

35. Ponschock, R.L,. & Becker, G.F., (2011). CLOUD Technology: A transformational dynasty on the ICT evolutionary continuum and contemporaneously a societal speciation event. *European Journal of Management.* ISBN: 1555-4015.

36. Ponschock, R.L., & Greif, T. B. (2007). Archeological excavating in virtual villages: A primer on discovery of artifacts from a digital community. *Proceedings of the IABE 2007 Annual Conference, 3* (1), 260-265.

37. Reddiar, C., Kleyn, N., & Abratt, R. (2012, September). Director's perspectives on the meaning and dimensions of corporate reputation. *South African Journal of Business Management, 43*(3), 29-39

38. Resnick, P., Zeckhauser, R., Friedman, K., & Kuwabara, K. (2000). Reputation Systems: Facilitating trust in Internet transactions. *Communications of the ACM, 43*(12), 45-48

39. Rhee, M. & Valdez, M.E. (2008). Contextual factors surrounding reputation damage with potential

implications for reputation repair. *Academy of Management Review, 34*(1), 146-168.

40. Rodriguez, T. (2013, October). Six steps for managing your online reputation. *AAOS Now, 6 (10), 29-29.*

41. Rose, B. (2011). STAKING your Reputation. *ABA Journal, 97*(4), 48-53.

42. Shandwick, W., (2009). *Risky business: Reputations online.* Retrieved May 15, 2013 from http://www.online-reputations.com/

43. Scharmer, C. O. (2007). *Theory U: Leading from the future as it emerges.* Cambridge. MA: Society for Organizational Learning.

44. Shiller, K., (2010, November). Getting a grip on reputation. *Information Today, 27 (10).*

45. Solve, D. J., (2004). *The digital person: Technology and privacy in the information age.* New York, NY. New York University Press.

46. Solve, D. J., (2007). *The future of reputation: Gossip, rumor, and privacy on the Internet.* New Haven, CT: Yale University Press.

47. Stegmeir, M. (2012, Jun 2). *Nancy Sebring's sexually explicit e-mails disclosed.* Retrieved August 1, 2013 from
http://pqasb.pqarchiver.com/desmoinesregister/doc /1019427585.html?FMT=ABS&FMTS=ABS:FT&type =current&date=Jun+2%2C+2012&author=STEGMEI R%2C+MARY&pub=Des+Moines+Register&edition =&startpage=&desc=Nancy+Sebring%27s+sexually+ explicit+e-mails+disclosed

48. Svantesson, D. (2009). The right of reputation in the Internet era. *International Review of Law, Computers & Technology, 23*(3), 169-177.

49. Tomkins, P., & Lawley, J. (2005). *Feedback loops.* Retrieved April 28, 2013 from http://www.cleanlanguage.co.uk/articles/articles/22 7/2/Feedback-loops/Page2.html

50. Warmoth, A. (2000). *Social constructionist epistemology.* Retrieved May 21, 2012, from http://www.sonoma.edu/users/w/warmotha/episte mology.html

51. Warren, S. D., & Brandeis, L.D. (1890). *The right to privacy. Harvard Law Review,* 4 (5), 193-220

52. Woodward, D. (2010). Reputations at RISK. *Director (00123242), 63*(10), 56.

Chapter 8: Technological Society

This chapter expands on the theoretical contexts of a speciation event related to the digitization interrelationship among the business, academic and political societal norms. Based on the prior research [42], it is evident that existent theoretical constructs among these distinct environments are increasingly inter-dependent as related to the velocity of change in social networking capabilities. The further research inherent within is formed from a meta-analysis of existent theory, and ultimately tied to the increasingly avant-garde technological advances that bring these global societal groups into a coherent model of interconnectedness.

Further research is still required to more closely bind these elemental aspects and forge a holistic view for positive sustainability and leverage among these groups.

Background

Industry, education, politics, are inseparable; combined they encapsulate the content of the global society. Therefore, when speaking of one construct we are speaking to all three.

C. Otto Scharmer advanced the thought that the world society is in a period of genesis [48]. The sociologist Manuel Castells of the University of California Berkeley in his book "End of Millennium", argued that our current societal environment was formed in the late 1960s and early 1970s, from among other drivers the information technology revolution and a new culture of "real virtuality". We refer to this environment as the digitization of society. If society continues on the continuum of the Information Communication Technology (ICT) dynasty it will reach a transformation or even more dramatic a societal speciation event [42].

C. Otto Scharmer (2007) in his book *Theory U* described transformation as the "U process". Scharmer's "U process" submits that "transformation must pull us into an emerging possibility and allows us to operate from that altered state rather than simply reflecting on and reacting to past experiences".[48,p5] The current state in the digitization of our society is beyond a point of retraction. Kuhn referred to a transformation of this magnitude as a "Paradigm shift".[32]

A century ago, social interactions involved associations with others who were within a walking radius[13]. For many, especially in the industrialized West, small face-to-face

167

communities are disappearing. Technology and the Internet have introduced communities that do not exist in geography and have no tangible physical presence. These virtual villages or townships[43] are not represented by geography, social class, or financial accounting. Instead, their cyber position is defined and driven by curiosity.[33] As Laurie Anderson musician/artist wrote "Technology is the campfire around which we gather". Social constructionists argue that the authority of knowledge ultimately derives from knowledge communities who agree about the truth.[46]

As Thomas Kuhn states in *The Structure of Scientific Revolutions* "Knowledge is intrinsically the common property of a group or else nothing at all."[32] The legacies of 845 million Facebook subscribers[22], not counting the additional millions of people using the other social networking communities like Twitter, and Myspace, are being archived for future societies to examine. The citizens of cybernetic civilizations are now being buried in the form of personal likes, dislikes, dreams, and possibly "dirty laundry" in landfills of virtual villages or virtual communities.[9]

Berger and Luckman advanced the theory of social construction when they noted that everyday life is "not only taken for granted as reality by the ordinary members of society in the subjectivity meaningful conduct of their lives, it is a world that originates in their thoughts and actions and is maintained as real."[6] A shift is occurring in the way members of our technologically oriented society interrelate[37]. It is changing from one of personal relationships between individuals to virtual chats. Technology has been driven by the "new Media" over the past 10 years. A distinguishing attribute of a true technological reconstruction is that many innovations occur at about the same time.[31]

In keeping with these types of transformations, Deutschman in his study of change reported three major factors involved in making a positive difference in increasing people's mastery and the ability to change: reframing change, radical change, and multifaceted support of change. He found that the way change is framed is important. [11] The reframed message must be positive and inspiring. How change propositions are framed can scare people off [21] or inspire and sustain them. Essentially mastering transformative change is necessary because time and resources are too limited to keep revising transformational change processes without success, especially given the rapid fire pace of technological innovation.

This chapter will focus on the combination of the CLOUD and social medial. Dortch refers to this combination as "the mobile social cloud" [12] and advanced the thought that with "the mobile social cloud every customer, partner, prospect and pundit is an actual or potential influencer of your customers and prospects."[12]

This section will systematically deliver dialog and provide a meta-analysis on the digitization affects that have already influenced the way society and it's actors function. Most of this is not news. The "end state" however is becoming clearer as to the need for a holistic model that provides for the positive leverage and sustainability of this speciation event. Bernam and Bell advanced the observation that all levels of society are using social networks to find jobs, restaurants, lost friends and new ones. Citizens, industrialists, academics, and politicians are delivering their message and brands through social networks. [7] The excerpt below from an IBM study on Digital Transformation illustrates the rapid shift toward digital evolution across all society functions.[7]

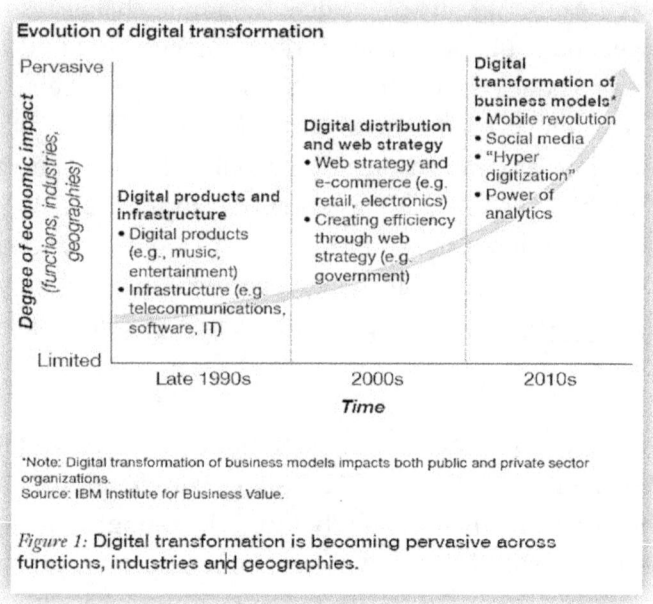

Evolution of digital transformation

*Note: Digital transformation of business models impacts both public and private sector organizations.
Source: IBM Institute for Business Value.

Figure 1: Digital transformation is becoming pervasive across functions, industries and geographies.

Figure 24[1]: Evolution of Digital Transformation

Digitization of the Industrial Landscape

"In today's increasingly digital world, even companies in primarily physical industries will start their digital transformation..."[7] With the goal of improving decision cycle times and organizational effectiveness, companies are utilizing social media solutions such as micro blogging, social media, wikis, and internal blogs. Social media is an online environment created for the purpose of mass collaboration.

The Aberdeen group, a business research organization, stated in their 2011 business review that 43% of organizations reported that they will invest in external social media in the next 12 months. The same study also points to the use of mobile apps as a one of the significant drivers over the next year. According to the Aberdeen study, enterprise mobility incorporates two fundamental components: 1) Management of the end-to-end mobile eco system, and 2) Enterprise class mobile software. [40] The mobile eco system has fueled the latest disruptive technological paradigm shift to *Bring Your Own Device* (BYOD). BYOD is a shift from using only company provided technology for business tasks and only connecting specific equipment to company networks to permitting employee to use their own tablets, cell phones, and other device for company business. "The explosion of mobile devices has been both a boon and bane to enterprises. ... The proliferation of BYOD will not slow anytime soon. Gartner another research organization predicts 90 percent of organizations will support enterprise applications and consumer devices by 2014".[16] According to the *Computerworld Consumerization of IT* study, published October 2011, about ½ of the 604 respondents said their organizations allow employees to do work using their own devices.[27] Monica Basso, a research vice president at Gartner indicated that previous business landscape supported collaboration through e-mail and highly structured software only. She concludes with the thought "Today, social paradigms are converging with e-mail, instant messaging (IM) and presence, creating new collaboration styles." ... "Technology is only an enabler; culture is a must for success."[25,28]

While social media is reshaping enterprise communications, other business communications are also evolving. Newer employees will enter the workforce with a bias to communicate via the new media, they will use e-mail in parallel to adhere to corporate protocols. Ms. Basso indicates "The rigid distinction between e-mail and social networks will erode. E-mail will take on many social attributes, such as contact brokering, while social networks will develop richer e-mail capabilities." Social media, the new media, is an online environment created for the purpose of mass collaboration[3]. For example, Facebook is a social media environment built on social networking technology, and Wikipedia is a social media environment built on wiki technology.[8] Gartner forecasts the new media is trending to replace e-mail as the principal tool for social communications for 20 percent of business users by 2014. [16]

According to a new study by IDG Research, which surveyed more than 260 large-enterprises IT managers, the vast majority of knowledge workers (86 percent) placed a very high level of importance on collaborating with internal coworkers and external stakeholders, and having access to the most up-to-date corporate information.[27] While AMI-partners has recently estimated that there are 3.5 million mid-sized businesses (SBMs), U.S. businesses uses smart phones as a corporate tool. They predict this to rise by 40% during 2012.[12]

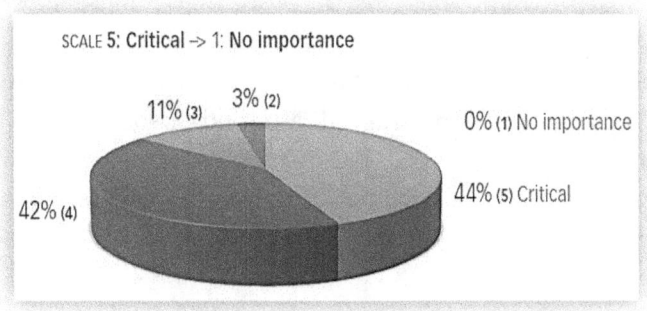

Figure 25: Collaboration Indicators[27]

The chart was derived from a survey, in which 86 percent of respondents said that having the ability to collaborate securely both within and beyond organizational boundaries is critical or very important. Additionally, 55 percent said that having access to the latest technologies (cloud, mobile and socially enabled) is an important component to achieving that goal.[27]

Social networks are integrating employees, customers, and suppliers that were not even contemplated five to ten years ago. As Prahalad & Krishnan note "There is a fundamental transformation of business underway. Forged by digitization, ubiquitous connectivity, and globalization, this will radically alter the very nature of the firm and how it creates value..."[44] Partnerships between customers and suppliers are morphing into global, complex, interdependent exchange points, forcing organizations to extend planning beyond the four walls of their enterprise. [20] Many firms find themselves unprepared to accept the challenges posed by this new reality as managers and leaders face deeply engrained organizational legacies steeped in both social and technological agendas.[44] Management in the 21st century plays a pivotal role in transforming mindsets, skills, behaviors, and decision structures of organizational leadership. This transformation to a virtual social architecture will only be successful through a period of socialization and acculturation where the organization and its component parts work together organically to effectuate transformative change.

Digitization – The Social Collective

The power of crowds can be observed from nonviolent "Black Friday" gift buying to "flash mob" chaos. "Police departments in several cities around the country are investigating what appear to be incidents of "flash mob"-generated violence, in which packs of dozens or even hundreds of youths appear seemingly out of nowhere to commit assaults, robberies and other crimes against innocent bystanders."[15] Technology-enabled collaboration can bring together groups/ "mobs" of people through texting, group e-mails, "tweets" and other social media exchanges.

Although frustrations generated by the pains of the economic recession fed the "Occupy" movement, "its conduit to the masses (and each other) is social media— and Twitter in particular"[41]; "Social media, a main engine driving "Occupy Wall Street's" spread, both nationally and globally… ."[19] Cyber-chatter is viral and can spread rapidly. Every day, millions of tweets, posts, and texts swell the global collaboration. Crowd mentality, technology-enabled collaboration draws us closer, makes us smarter and allows us to innovate through the wisdom

174

of a crowd. A new wave of collaborative consumption is transforming consumerism and the rules of engagement .[30]

The viral nature of the new social media played significantly in the 'Arab Spring' in the Middle East region; the new media appeared while traditional modes of communication and civil freedoms were suppressed and restricted. Facebook and YouTube are obviously essential parts of this new news ecosystem.[24] With the new social media emerged "citizen journalists" -- on the street average people with cell-phone cameras. The new media illuminated a social mobilization signaling a new paradigm or even a synthesis of new digital media and conventional communication channels.[39]

The new media has also entered the United Stated political landscape. Howard Dean, the former aspirant for the presidency proved that the web could be a force for raising money online. Barack Obama trumped Dean with an unimagined online contribution level. Obama collected nearly a half a billion dollars from online contributions; attracting more than 3 million contributors. Social media played a critical role in organizing support for Obama's historic election.[45]

After the election Obama continue relying on the new media with the presentation of his weekly presidential radio address through YouTube. Obama continues to prove the power of social networks. Riyaad Minty a spokesman for Al Jazeera's summarized the new media as "a giant speech bubble for what's happening in the world."[24]

Digitization in Personal Lives

The following graphic illustrates that there are 179.7 million social media users recorded in 2015. The internet and social networking has become the means of social interaction possible replacing discussions on the street corner, coffee shop, local pub, or barber chair.

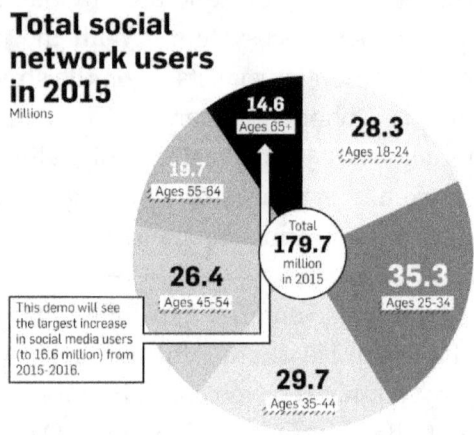

Total social network users in 2015
Millions

14.6 Ages 65+
28.3 Ages 18-24
19.7 Ages 55-64
Total 179.7 million in 2015
26.4 Ages 45-54
35.3 Ages 25-34
29.7 Ages 35-44

This demo will see the largest increase in social media users (to 16.6 million) from 2015-2016.

Figure 26: Global Internet Use

Global Positioning Systems

Like the Internet, Global Positioning Systems (GPS) have become an essential component of the global information dependent society. The free, open, and dependable characteristics of GPS have led to hundreds of uses affecting every facet of 21st century culture. GPS technology is now in everyday items like cell phones to farm tractors, shipping containers, and automobiles. The GPS is a space-based navigation system that utilizes a triangulation of signals from several satellites (four or more at one time) that provides location and time information anywhere on the globe. The GPS program provides critical capabilities to military, civil and commercial users around the world.

General Motors (GM) first introduced onboard navigation in automobiles with the introduction of OnStar. OnStar advocates it as an essential safety tool. The benefits, they say, include its ability to aid police in tracking down stolen vehicles, contacting emergency medical services in case of an accident, and notifying drivers of potentially dangerous mechanical problems. Owners are e-mailed the results of remote diagnostic. Some of these services or apps are free while other features are part of a subscription service. Similar services are now offered by Ford. Ford's SYNC® technology adds the Voice recognition / hands-free option. Through voice command the systems not only directs the driver when to turn but can find and play his/her favorite music.

Precision Agriculture

Precision agriculture is changing the way farmers and agribusinesses view the land from which they reap their profits and in turn feed the world's population with a low cost, safe food supply. Precision agriculture is about collecting timely geospatial data on soil-plant-animal requirements and prescribing and applying location targeted pest and plant management treatments to increase agricultural production and contemporaneously protecting the environment. Precision agriculture has been made possible by integrating the Global Positioning System (GPS) and geographic information systems (GIS). These technologies combine real-time data retrieval with accurate positioning. GPS allows farmers to work when visibility of the field is hampered by darkness, rain, dust, or fog.[18]

> *We started using more digital technology in the last 10 years. We have gone to GPS (Global Positioning System) for a handful of different operations from cultivating to planting. By using GPS on the tractors, the entire process from leveling the field to planting the seed to irrigating the crop has been much more efficient than in the past. GPS is used in a lot of applications throughout most aspects of agriculture. John Boelts, Vice President of the Yuma County Farm Bureau; Yuma, AZ*

Car and Driver

Google has launched the driverless car. "While this project is very much in the experimental stage, it provides a glimpse of what transportation might look like in the future; thanks to advanced computer science."[34] Engineers told the New York Time that the forays onto the highways have been largely incident-free, apart from one bump when the car was reportedly hit from behind at a traffic light. Driverless cars are years from mass production, but technologists who have long dreamed of them believe that they can transform society as profoundly as the Internet.[34]

The World Health Organization has shown that more than 1.2 million people are killed each year on the roads. Mr Thrun - professor of computer science and electrical engineering at Stanford University said that number could and should be reduced with the driverless technology. "We believe our technology has the potential to cut that number, perhaps by as much as half."[4]

Self-parking (parking assist) cars, backup obstruction braking, lane drifting has advanced rapidly. Driverless technology is a union of existing capabilities. In addition to the safety advances the technology will bring to transportation; once perfected the moving of goods and current supply chains will be dramatically transformed.

Look to the sky

A few decades ago many children's books and even some television advertisements showed storks delivering babies.

Image 3:

Today amazon, FedEx, and others are experimenting with similar Drone delivery options. The major differences are the "beak", the blanket, and the nine-month cycle time.

Image 4:

Idol Time

Cellphones, computers, and text messages overflowed on May 15, 2012. The battle for the American Idol title between finalists Philip Philips and Jessica Sanchez raised cyber chatter to a record level. American Idol" viewers cast a record 132 million votes in the showdown between bluesy guitar man Phillip Phillips and teenage songbird Jessica Sanchez, host Ryan Seacrest said during Wednesday's finale. Astonishing this compares to between 126.5 million and 128.5 million Americans cast ballots in the 2008 presidential election according to a report by the electorate study center.[29] From "Idol" came other interactive audience participation results shows, the VOICE and America's Got Talent, among others.

Digitization of Education

Although brick and mortar academia is not going away anytime soon, a trend line to "online" or eLearning is showing continual upward signs. The number of students taking at least one online course continues to expand at a rate outpacing the rate of traditional higher education enrollments. The most recent estimate, for 2013, places this number at 5.2 million online students, an increase of 9.3

percent over the previous year [1] even though the growth rate is declining.

GROWTH RATE OF NUMBER OF STUDENTS TAKING AT LEAST ONE ONLINE/DISTANCE COURSE – 2003 TO 2013	
Fall 2003	23.0%
Fall 2004	18.2%
Fall 2005	36.5%
Fall 2006	9.7%
Fall 2007	12.9%
Fall 2008	16.9%
Fall 2009	21.1%
Fall 2010	10.1%
Fall 2011	9.3%
Fall 2012	6.1%
Fall 2013	3.7%

Figure 27: Online Learning Consortium Growth Rate[1]

The world is evolving beyond what was considered traditional. In every facet of life, we are seeing an acceleration of learning, success and self-improvement that outstrips the very idea of a four-year college degree. Young adults two years out of high school are becoming millionaires using the strength of their intellect and determination.[2]

Online degrees are widely accepted. "Predictably, the best numbers from the Vault study were logged in Internet/New Media (70% projected acceptance), Technology (46%), High Tech (44%), and Marketing/Media (29%). Areas like Medicine and Law were described as the least likely to accept online credentials."[38] In a recent Online consortium survey 57.9% indicate that online outcomes are the same as face-to-face.[1] Institutions are engaged in online learning. Online offerings have become a strategic component of academics.

ONLINE EDUCATION IS CRITICAL TO THE LONG-TERM STRATEGY BY INSTITUTIONAL CONTROL – 2006 TO 2014		
	Public	Private nonprofit	Private for-profit
Fall 2006	74.1%	48.6%	49.5%
Fall 2007	70.7%	47.1%	53.2%
Fall 2009	73.6%	49.5%	50.7%
Fall 2010	74.9%	52.3%	60.5%
Fall 2011	77.0%	54.2%	69.1%
Fall 2012	77.3%	65.1%	61.3%
Fall 2013	73.6%	63.8%	54.9%
Fall 2014	72.9%	64.5%	80.9%

Figure 28: Online Learning Consortium Profit Status[1]

Digital Transformation

Given the various elemental digitization events that have transpired and continue to gain velocity, it appears evident that a holistic interrelationship between the global sectors is upon us, especially related to the business, academic and political landscapes. Similar models and tools are being leveraged within these sectors, thereby breaking down any independent leverage; rather, the leverage remains inherent across these historically disparate communities and sectors. Horrigan posits that these technological advances will continue at an accelerating pace, thereby inhibiting individual leverage.[23]

Consider that theoretical constructs dating 20-30 years were a forbearer of these technological innovations and convergence. Beninger discussed these very advances as having the potential for unique leverage as well as a societal convergence.[5] Therefore, the technological advances are a speciation event formed and evolving to further the convergence of global sectors resulting in sustainable and positive influence.

Chapter Wrap-up

The digitization of life in the 21st century has infiltrated our very existence, only a few of which were examined in the paper's content. One of the transformation events in the 21st century is social networking which the CLOUD is both the ends and the means. Social networking is a powerful force throughout the world. We've recently seen the power of Twitter and other microblogging tools in Egypt and Libya. The entire world uses these tools everywhere. Childers adds "Twitter and Facebook are fantastic at what they do. Connecting you with your friends, but look forward to the next generation of niche social networking sites, centered on every type of interest imaginable."[10]

This discussion is not about the technology but rather the transformational impact it has placed on society in general and its contribution to the speciation of the individual. We also contend that the interrelationship between business, academic and political sectors is rapidly converging based on the leveragability of this societal speciation event.

Beniger, a seminal thinker on society and evolutionary control posits

As in earlier revolutions in matter and energy technologies, the nineteenth-century revolution in information technology was predicated on, if not directly caused by, social changes associated with earlier innovations; Because technology defines the limits on what a society can do, technological innovation might be expected to be a major impetus to social change.[5]

Social networking enabled by the Internet is performing crucial roles throughout society. Just a few of the positive purposes social networks can be used for include:

- Researchers can collaborate between any location in the world
- Friends/families are no longer bound by distance
- Academics can use Social networks to enhance courses and knowledge transfer
- Businesses can use social networks internally and along the supply chain

Digitization and social networking has migrated into businesses of all sizes as a way to obtain secure access to advanced technology that is not necessarily owned or hosted by the user.[25] Outside of the boundaries of industry, some 69% of online Americans use webmail services, store data online, or use software programs such as word processing applications whose functionality is located on the Web. In doing so, these users are making use of "cloud computing," an emerging architecture by which data and applications reside in cyberspace, allowing users to access them through any web-connected device.[23]

Work life and the time spent business tasks with the use of mobile devices have blurred the boundaries between personal and work events. Online education has reached deep into the roots of higher education and now even available at the high school level. Personal collaboration intersects education and business, while politics has become more than just discussion around the office. None of these measures happened overnight.

Through social constructionism individual behaviors and group participation in the building of perceived social realism can be uncovered. Digitization of society and "Internet-based, just-in-time delivery of data, applications, storage, and computing power as services are done in a way that completely shields consumers from underlying technical details."[36] Digitization in general and social networking specifically, have profoundly changed and continues to fuel social change. On account of traditional Darwinian account of organic evolution, variation is supplied primarily by mutations which are random relative to the potential beneficial contribution to fitness of the modified trait. "Evolutionary theory applies to *populations* of individuals."[17]

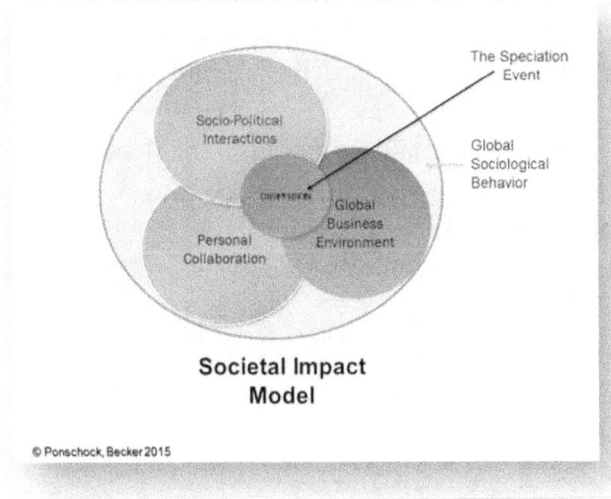

© Ponschock, Becker 2015

Figure 29: A Convergence Model

185

The evolution of new species doesn't usually happen in an afternoon, so it's quite difficult to see. And, on the geological scale, mankind has been around for a blink of an eye. Still, we have observed speciation and see its telltale fingerprints everywhere. Evolutionary theory is a very intellectually powerful concept indeed. It has pushed biology to the fore of science and it has given answers to questions that would have otherwise gone unanswered or simply put down to an infamous deity's 'mysterious ways'. Mankind has always striven to understand life, the universe and everything; the theory of evolution goes a long, long way to reach this lofty, ambitious goal.[14]

The relationships depicted in the "convergence model" (Figure 29) illustrate the junction of previously bounded and diverse facets of societal life becoming a single new entity. The four domains previously held beyond arm's length are melded into a digital environment creating an atmosphere that will permanently change society and the way humans interact within and between cultures.

Continued observation of the digital markers that will trigger the intermediate and final speciation events is paramount. The collection and mining of massive amounts of data is one of the drivers being reviewed in the next chapter. This data growth is referred to in the information technology sector as "Big Data".

186

End Notes

1. Allen, I.E & Seaman J., (2015). Grade level: Tracking online education in the United States. *Sloan Consortium.*
2. Almeda, (2012). Traditional universities struggle to compete with online counterparts. Retrieved April 22, 2012, from http://almedauniversity.org/articles/article4.html
3. Baso, M. (2010, November). Analysts' discussion on future for mobile collaboration. *Gartner Symposium/ITxpo 2010.*
4. BBC (2010, Oct 10) *Google tests cars that drive themselves.* Retrieved May 20, 2012 from http://www.bbc.com/news/technology-11508351.
5. Beniger, J. (1986). *The control revolution: Technological and economic origins of the information society.* Harvard University Press Cambridge MA
6. Berger, P.L., & Luckman, T. (1967). *The social construction of reality: A treatise in the sociology of knowledge.* Garden City, NY: Doubleday.
7. Berman. S.J., & Bell. R. (2011). *Digital transformation: Creating new business models where digital meets physical.* Somers, New York: IBM Institute for Business Value.
8. Bradley, A. J., & McDonald, M.P., (2011). *Social media success is about purpose (not technology).* Retrieved May /2012, from http://blogs.hbr.org/cs/2011/11/social_media_succes s_is_about.html
9. Cashmore, P. (2006, August, 9). MySpace hits 100 million accounts: Mashable social networking 2.0.

Retrieved November 9, 2006, from http://mashable.com/2006/08/09/ myspace-hits-100-million-accounts/

10. Childers, A., (2011, Jan). *Niche social networks in a Facebook era.* Retrieved May 3, 2012, from http://www.adrianchilders.com/niche-social-networks-in-a-facebook-era/

11. Deutschman, A. (2005). Make change or die. *Fast Company, 94*(5), 52-60.

12. Dortch, M., (2012). *Dortch on SaaS & cloud computing: The blog.* Retrieved May 21, 2012, from http://dortchonsaas.blogspot.com/2011/11/inventory-management-and-mobile-social.html

13. Ermann, M. D., Williams, M. B., & Shauf, M. S. (1997). *Computers, ethics and society.* New York: Oxford University Press.

14. Evolution (2002). The theory of evolution – part II. Retrieved May 1, 2012, from http://h2g2.com/dna/h2g2/alabaster/A737985

15. Fox News, (2011). *Questions arise over whether 'flashmob' attacks in U.S. Cities motivated by race.* Retrieved May 20, 2012 from http://www.foxnews.com/us/2011/08/10/flashmob-attacks-in-us-cities-raise-questions-over-possible-race-motivation/#ixzz1vblyxhWJ

16. Gartner's top predictions for IT organizations and users, 2011 and beyond: IT's Growing transparency.

17. Giere, R. (1999). *Science without laws.* Chicago, IL: University of Chicago Press.

18. GPS Applications retrieved May 5, 2012, from http://www.gps.gov/applications/

19. Goodale, G. (2011). *Social media occupy Wall Street: Do they divulge its secrets.* Retrieved 5/27/2012 from http://www.csmonitor.com/USA/Politics/2011/1012 /Social-media-drive-Occupy-Wall-Street.-Do-they-also-divulge-its-secrets
20. Harrington, L. (2007, April). Defining technology trends. Inbound Logistics. Retrieved May 15, 2008, from http://www.inboundlogistics.com/articles/features/ 0407_feature01.shtml
21. Hammond, J.S., Keeney, R.L. & Raiffa, H. (1998). The perfect decision. *Inc., 20*(14) 74-78.
22. Henschen, D., (2012, February, 27). From CRM to Social. *Informationweek.*
23. Horrigan, J. A., (2008, September). *Use of cloud computing applications and services.* Pew Internet and American Life Project.
24. Hounshell, B., (2011, July/Aug). *The revolution will be tweeted.* Retrieved 5/22/2012 from http://www.foreignpolicy.com/articles/2011/06/20/t he_revolution_will_be_tweeted.
25. *IDC finds marketers need to work on influencing buyers with social media* (2012, May) Retrieved May 18, 2012, from http://www.btobonline.com/apps/pbcs.dll/article?A ID=/20120516/SOCIAL06/305169994/idc-finds-tech-marketers-need-to-work-on-influencing-buyers-with&template=printart
26. IDOL, (2012). *World record breaking 132.3 million votes were cast at American Idol.* Retrieved May 27, 2012, from http://www.theentertainmentlifestyle.com/2012/05/ world-record-breaking-1323m-votes-were.html

27. IDG. (October 2011) *Consumerization of IT.* Retrieved May 13, 2012, from http://www.idgenterprise.com/wp-content/uploads/2012/01/IDGE_CoIT.pdf

28. Idinopulos, M., (2011). *Six steps to drive social software adoption.* Retrieved May 12, 2012, from www.socialtext.com

29. Joyner, J. (2008). *2008 Voter Turnout Same as 2004.* Retrieved May 27, 2012, from http://www.outsidethebeltway.com/2008_voter_tur nout_same_as_2004_/

30. Khondker, H. H., (2011, Oct). Role of the New Media in the Arab Spring. *Globalizations, Vol. 8,* pp. 675-679.

31. Kranzberg, M. (1989). IT as revolution: The information age. In T. Forester (Ed.). *Computers in the humancontext* (pp. 19-32). Cambridge, MA: MIT Press.

32. Kuhn, T.S. (1996). *Structure of scientific revolutions.* Chicago: University of Chicago Press.

33. Luthra, N. (2006). *The "Real" and the "Virtual" in public space.* (Master Thesis, University of New York at Buffalo). (UMI 1431955).

34. Markoff, J., (2010). *Google cars drive themselves, in traffic.* Retrieved April 3, 2012, from http://www.nytimes.com/2010/10/10/science/10goo gle.html?_r=1&%20r=2&src=tptw

35. McDonald, M., & Aron, D. (2011). *Reimaging IT: The 2011 CIO Agenda.* Gartner Executive programs.

36. Moser, M. (2009). *Workload automation: Helping cloud computing take flight.* Retrieved May 1, 2011, from http://documents.bmc.com/products/documents/39 /17/123917/123917.pdf

37. Mumford, L. (1970), *Myth of the machine: The pentagon of power.* New York: Harcourt.
38. *Online vs. traditional degree* (2011). Retrieved March 2, 2012, from www.collegeconfidential.com/distance/online-trad.htm.
39. Papandrea, M., (2007). Citizen journalism and reporter's privilege. *Boston College Law School Legal Studies Research Paper Series.* 110.
40. Park, H., (2012). *The 2011 Aberdeen Group business review.* Aberdeen Group: A Harte-Hanks Company
41. Politico (2011). *Wall Street will 'occupy' social media.* Retrieved May 1, 2012 from http://www.politico.com/news/stories/1111/67510.html#ixzz1vbq8DyWA
42. Ponschock, R.L,. & Becker, G.F., (2011). CLOUD Technology: A Transformational Dynasty on the ICT evolutionary continuum and contemporaneously a societal speciation event. *European Journal of Management.* ISBN: 1555-4015.
43. Ponschock, R., & Greif, T. B. (2007). Archeological excavating in virtual villages: A primer on discovery of artifacts from a digital community. *Proceedings of the IABE 2007 Annual Conference, 3 (1),* 260-265.
44. Prahalad, C.K. & Krishnan, M.S. (2008). *The new age of innovation: Driving co-created value through global networks.* New York: McGraw-Hill.
45. Vargus, J. A., (2008). *Obama raised half a billion online.* Retrieved April 25, 2012, from http://voices.washingtonpost.com/44/2008/11/20/obama_raised_half_a_billion_on.html

46. Warmoth, A. (2000). *Social constructionist epistemology.* Retrieved May 21, 2012, from http://www.sonoma.edu/users/w/warmotha/episte mology.html
47. Henschen, D. (2012, Feb 21). From CRM to Social. *Informationweek.*
48. Scharmer, C. O. (2007). *Theory U: Leading from the future as it emerges.* Cambridge. MA: Society for Organizational Learning.

Chapter 9: Datafication

On March 27ᵗʰ, 2014, a group of major Information Technology and equipment companies banded together to form a singular working group in an effort to promote access to Big Data and the Internet of Things. As many as ten companies, formed the Industrial Internet Consortium [6] in an effort to "drive standards for the so-called Internet of Things." [6] They hoped to do away with technological guards to enable enhanced Big Data accessibility and encourage incorporation of both the physical and digital environments. [13]

The Consortium wants to implement the accessibility by outlining an architectural structure for open production standards that would work with a wide range of market sectors from automotive and manufacturing to healthcare and the military, [9] and in the process supply best practices, reference architectures, case studies and standards requirements to make distribution of connected technologies much easier. [7]

193

Background

"The Internet of Things is a technological revolution in the future of ICT that is based on the concept of anytime, anyplace connectivity for anything."[14] This technology allows the ability to remotely monitor and adjust everything from our home thermostats to the electrical power grids of a major city, and the timely delivery of goods and services. We will learn how the Internet of Things has already changed our daily lives and how much more is still planned for the future.

The Internet of Things (IoT) is revolutionary technology that is renovating our society in almost every aspect. Individuals and businesses have been enjoying the steady benefits of IoT since the early 90s; however, recent hardware, software, and Big Data concepts have revived it. In the past, IoT technology was limited to the number and intelligence of devices. Now, IoT technology advancements can connect countless devices to the Internet, ranging from water meters, delivery fleet vehicles, and medical devices. More importantly, proactive monitoring, reporting, and modifications can now be done in real-time from almost anywhere in the world.

What is this So-Called "Internet of Things?"

In order to better understand the significance of the Industrial Internet Consortium's announcement, we must first understand this so-called "Internet of Things". Margaret Rouse, a writer for WhatIs.com – TechTarget's Information Technology encyclopedia and learning center - defined the Internet of Things as "a scenario in which objects ... are provided with unique identifiers and the ability to automatically transfer data over a network without requiring human-to-human or human-to-computer interaction."[10] She also believed that Internet of Things had evolved from the union of wireless technologies, micro-electromechanical systems – also known as MEMS, and the Internet.[10] The *thing* in this case can be a person with a heart monitor implant, a family pet – like a dog – that is chipped for identification purposes, a vehicle with built-in GPS capabilities and sensors, or any item that can be allocated an IP address and provided with the capability to transfer and access data over a network. [10]

The term "Internet of Things" was first coined in 1999 by Kevin Ashton, co-founder of the Auto-ID Center at the Massachusetts Institute of Technology. He used the phrase to describe an arrangement where the Internet is attached to the physical world by way of universal sensors.[1] Ashton believed in 1999 that "today's computers – and therefore the Internet – are almost wholly dependent on human beings for information." [1] He restated this belief in 2009, adding that nearly all of the approximately fifty petabytes of data available on the Internet was first captured and created by human beings. [1] To fully realize how much data fifty petabytes truly is, one petabytes is equal to 1,024 terabytes.

The Industrial Internet Consortium wants to be the leader in instituting interoperability across many different industrial environments for a more interconnected world. They believe this will help organizations connect and improve their properties, processes, and data more effortlessly and with more agility, with the expected result to be unlocked business value across all sectors and greater adoption of industrial Internet applications, which they believe is a starting position for speeding up the reality of the Internet of Things. [7] Ron Ambrosio, Chief Technology Officer for Smarter Energy Research at IBM, stated that "IBM's vision of a smarter planet is being realized as we connect more of the physical world with the Internet, pairing the Internet of Things with advances in analytics, mobile and cloud computing in ways that lead to new insights and efficiencies that can be harnessed for competitive advantage. Smarter cities, utility grids, buildings and machines are becoming more instrumented, interconnected and intelligent, and through this consortium, we will accelerate both innovation and technology advancement."[7]

The Industrial Internet Consortium will work together to develop what they call a 'common blueprint' that equipment and devices from all manufacturers can use to network together. They do not want these standards to be restricted to just Internet protocols, but to also include metrics like data storage capacity in information technology systems, the power levels within connected and non-connected machines, and data traffic control. The Industrial Internet Consortium will focus primarily on applications related to industries like healthcare, manufacturing, oil and gas exploration, and transportation [13] because quite often in these industries, hardware and software products are not compatible with one another.

A growing number of international companies and organizations wish to join and cooperate with the Industrial Internet Consortium, and with good reason.

Wikibob's Jeff Kelly, speaking with GE's CEO Jeff Immelt at 2013's D-Eleven conference, stated "there's a huge opportunity for those vendors who can help power plants or water treatment facilities ... to help them become more efficient and intelligently interconnect all these devices. In terms of where the [action is going to] be in the next five years in Big Data, [it is] absolutely going to be around Industrial Internet."[12] Furthermore, Richard Soley, the Industrial Internet Consortium's first executive director says that international organizations such as Fujitsu and Siemens have also shown interest in joining the group. He also believes that, had the consortium been formed five years earlier, the mystery surrounding the disappearance of Malaysian Airlines Flight 730 would not be an issue. Soley went on further by saying that if the airline "had full interoperability with the world's tracking systems, [we would] know where it is to a square meter."[13]

However, there is potential for serious security issues with the consortium. David S. Levine, an affiliate scholar at the Center for Internet and Society at Stanford Law School, believed that if organizations wish to join and cooperate with the Industrial Internet Consortium, they must be willing to be a lot less secret with their assets. He said, "If the [Industrial Internet Consortium's] goal of breaking down technology silo barriers to drive better Big Data access and improved integration of the physical and digital worlds is to occur, then the IIC and its members will have to come to terms with their goals' impact on what they do with their most valuable secrets."[8] Levine believed that the risk of getting this balancing act wrong is significant; the consumer's willingness to purchase and utilize the technology could be affected. Lastly, he states that users will want to know what the Consortium knows about them and what it can do with that knowledge, if not in specifics, then at least in generalization.[8]

The Internals of the Internet of Things

According to a recent white paper published by IBM, the Internet of Things has three major dimensions. This three-tier model includes "components, building blocks, and system of systems."[14] Each module serves a specific and distinct purpose; the removal of one part causes the other modules to fail and the three-tier support model to come crashing down.

The components of the Internet of Things make up the foundational structures of the system. Components are specific to the application being supported; they serve a purpose in meeting a need. For instance, "a water system uses meters, pressure and flow sensors, and value control components" to fulfill a solution.[14] Without the specific components (meters, pressure and flow sensors) a water system would fail to function and/or serve a useful purpose.

The building blocks of IoT serve as the backbone and are critical to the success of solutions. They can consist of, "communication, security, analytic engines, remote computer nodes, and update engines."[14] Building blocks serve to create and maintain the systems we use in our everyday lives, without them life would be considerably more difficult.

Building blocks define the individual systems that are needed to support the solution. Individual systems combine to create the third module of IoT, system of systems. The grouping of various systems and its operational purpose makes the distinction in IoT. Let's use the healthcare industry to further illustrate the concept of IoT's system of systems. EMR, imaging, and medical devices/systems can all be utilized at the same time to administer treatment to patients. They are used to adjust prescription dosages and frequencies, and recommend when to begin rehabilitation therapy after a specific treatment or surgery. Successfully automated treatments could potentially be used by doctors to build treatment plans for future patients that share a similar physical, medical, and personal background to help them fully and safely recover.

The healthcare illustration used earlier can symbolize the importance of the individual components and systems used by the Internet of Things. Each system has its own level of autonomy, dependence, and interaction with the other parts. Without the other modules, the system of systems could not be combined together to produce a solution to a problem or improve a process.

Big Data & Internet of Things working together

The benefits gained from the Internet of Things technology have many corporate sectors considering the implementation of such systems. The ability for companies to instantly monitor processes, inventory levels, and coordinate adjustments to production procedures all in real-time is very appealing and beneficial to their businesses. The amount of data being generated by IoT hardware and software can be easily handled by Big Data applications.

In order for organizations to benefit from the capabilities of IoT they will need to invest heavily in IT services. They will inquire substantial costs for outside expertise, network hardware, Big Data applications and storage, training, and security. Organizations will need to ensure that they have a network capable of handling the additional amount of traffic from IoT- capable devices and the IT staff to support the entire infrastructure.

How IoT and Big Data can benefit Supply Chain Management

The Internet of Things and Big Data technologies will have a significant impact on Supply Chain Management. IoT has the ability to bridge the communication gaps that exist between various corporate industries that are very reliant upon Supply Chain Management in order to function because of supply and demand requirements. This communication gap has been maintained primarily because of older traditional database applications and structured systems that were commonly used in the past for Supply Chain Management to conduct their day-to-day business needs.

In the past, supply chain management relied heavily upon structured databases such as Oracle, SQL and SAP to conduct their business. These traditional database applications were common in conducting the supply and demand side of the business. However, they don't provide the real-time capabilities that unstructured/Big Data applications can offer. Instead of waiting days or weeks for more needed products or having excess merchandises, warehouses can now maintain an accurate number of items, thereby improving logistics and shelf life while reducing waste and expenses experienced because of a surplus. Or, as better stated by Michael Burkett, the managing vice president at Gartner, "As the number of software-embedded digital-physical products grows, the methods of product development and life cycle management across the supply chain will change."[15]

IoT technology can autonomously manage anticipated demands for either an increase or decrease in specific merchandises. For example, if the seasonal changes call for an increase or decrease in specific commodities IoT technology can proactively manage these changes based on previous years of accumulated data. This automated process is only possible because IoT-capable devices can communicate directly with both private and public networks for the transfer of gathered data. This information is sent to huge data repositories where it is accumulated, processed, and analyzed by Big Data applications such as Hadoop. Hadoop can analyze this large amount of data in real-time and process it into meaningful information which can be sent promptly to e-mail receipts for decision making and increased efficiency.

What will the involvement of IoT in Supply Chain Management look like for individual consumers? Imagine going into a hardware store and being able to buy the needed tools and supplies every time without having to leave frustrated because of an out of stock item(s). Or going to the grocery store and being able to buy what you need at lower prices accompanied with longer shelf lives. All because of the implementation and use of IoT technology which has allowed, "Supply chains to deliver more differentiated service to customers more efficiently."[15]

In 1999, Cherry Murray pronounced a vision of the World's "electronic skin"

> *In the next century, planet Earth will don an electronic skin. It will use the Internet as a scaffold to support and transmit its sensations. This skin is already being stitched together. It consists of millions of embedded electronic measuring devices: thermostats, pressure gauges, pollution detectors, cameras, microphones, glucose sensors, EKGs, electroencephalographs. These will probe and monitor cities and endangered species, the atmosphere, our ships, highways and fleets of trucks, our conversations, our bodies-even our dreams.* [4]

Although the above may read like it came from a science fiction magazine, as far back as the early 1990s, hundreds of thousands of personal computers working in unison have already tackled complex computing problems. An example of what is often referred to as massively distributed computing is SETI. SETI is an acronym for the Search for Extra-Terrestrial Intelligence. SETI is an Internet-based public volunteer computing project employing the Berkeley Open Infrastructure for Network Computing (BOINC) software platform, hosted by the

Space Sciences Laboratory, at the University of California, Berkeley. Its purpose is to analyze radio signals, searching for signs of extraterrestrial intelligence. About 401,969 volunteers and 968,500 computers (hosts) worldwide make up the computing network. In the future, some scientists expect spontaneous computer networks to emerge, forming a "huge digital creature". [5]

How IoT and Big Data can help the Energy Industry

The Internet of Things is already impacting and changing the world's energy industries. Big Data applications have already added tremendous value to the energy industry; IoT will only strengthen the benefits achieved by Big Data. Former GE CEO Jeffrey Immelt sums up the importance of IoT and Big Data by declaring "Data analytics and the study of the information spinning off the machines GE sells has become just as pivotal to industrial manufacturers as metallurgy and the study of the physical elements was in the last century."[16] The significance of Mr. Immelt's statement equates to massive savings for the energy industries and overall lower prices for consumers.

The total savings that energy companies can achieve from the successful implementation and use of Internet of Things and Big Data technologies is significant. To put this into perspective, GE released a report in 2012 detailing the power of the 1 percent. "1 percent improvements in healthcare, aviation, rail, power and oil and gas industries could bring $276 billion in efficiency savings to the global economy across the next 15 years."[16] Imagine what a five percent could do for the energy industries and the large manufacturing, transportation, and production industries that rely heavily upon energy for operations? These types of monetary savings and initiatives are becoming the reasons why so many industries are preparing to or increasing their utilization of the Internet of Things and Big Data technologies and applications.

Chapter Wrap-up

The innovations of the Internet of Things and Big Data technologies can help a variety of corporate industries within our economy. IoT will only continue to grow exponentially and IBM "predicts that by 2020, possibly 50 billion devices will be connected, a number that is 10 times that of all current Internet hosts, including connected phones."[14] With these impressive numbers more companies will plan to implement and expand their use of these two technologies, thereby interweaving various industries that will become greatly dependent upon one another.

The Internet of Things – and therefore the Industrial Internet Consortium - offers great possibilities in technological advancement. However, this may come at a price that consumers and corporations may be uncomfortable with. Industries may benefit from the ability to access archived data that will enable them to design safer, more efficient, and greener vehicles in the automotive industry, or to make more accurate diagnoses of patients in the medical industry. On the opposite end of the spectrum, corporations who get involved with the Consortium may have to be willing to be more open with their corporate trade secrets, which could violate consumer trust with the cooperating organizations if certain sensitive consumer information is exposed. The Consortium will have to maintain a balancing act between openness and privacy if it is to succeed.

This is a very basic and generalized example of how our lives are connected to big data. Big data is being integrated into most every facet of our lives. From the collection of information our modern cars are producing, to grocery shopping, and even our homes becoming "smart homes" with data being generated daily.

Information is being collected and analyzed by businesses to make better products, predict weather patterns, flu outbreaks, targeted advertising, etc. The uses for big data are seemingly endless as companies see the great value in data collection. As the access to digital content becomes more globally available, the increase in data output is going to be exponential. As we move forward as a global society, big data will become more and more integrated into our daily lives and routines. The continued growth of big data will allow companies to explore techniques to interpret, extract, and use stored information in ways never before imagined.

End Notes

1. Ashton, K. (2009). That 'Internet of things' thing. *RFID Journal*, Retrieved from http://www.rfidjournal.com/articles/view?4986
3. Coy, P., & Gross, N. 21 Ideas for the 21st Century. *Business Week Online* 30 August 1999.
4. Gross, N. (1999, August 30). 21 Ideas for the 21st Century. *Business Week Online* retrieved September 29, 2014 from http://www.businessweek.com/1999/99_35/b3644024.htm
5. Korpela, E., Werthimer, D., Anderson, D., Cobb, J. & Lebofsky, M. (2001). *SETI@HOME—massively distributed computing for SETI* retrieved September 18 2014 from http://setiathome.berkeley.edu/sahpapers/CISE.pdf
6. Lawson, S. (2014, March 27). Cross-industry IoT group pushes for gear that works together. *ComputerWorld*, Retrieved from http://www.computerworld.com/s/article/print/9247240/Cross_industry_IoT_group_pushes_for_gear_that_works_together (Lawson, 2014)
7. LeClaire, J. (2014, March 28). Cisco, IBM launch Internet of Things consortium. *Top Tech News*, Retrieved from http://www.toptechnews.com/story.xhtml?story_title=Cisco__IBM_Launch_Internet_of_Things_Consortium&story_id=011000WXZR7O (LeClaire, 2014)
8. Levine, D. S. (2014, April 4). What does the Internet of Things mean for corporate secrecy? *FutureTense*. Retrieved March 15, 2015 from http://www.slate.com/blogs/future_tense/2014/04/0

206

4/what_does_the_Internet_of_things_mean_for_cor
porate_secrecy.html (Levine, 2014)

9. Merritt, R. (2013, August 7). U.S. consortium
forming on industrial Internet. *EE Times*, Retrieved
from
http://www.eetimes.com/document.asp?doc_id=13
19162 (Merritt, 2013)Press

10. Rouse, M. (2013, July). *What is Internet of Things?*
Retrieved   from
http://whatis.techtarget.com/definition/Internet-of-
Things (Rouse, 2013)

13. Wheatley, M. (2014, April 1). IBM, Cisco, GE &
AT&T   form   industrial   Internet   consortium.
SiliconANGLE,   Retrieved   from
http://siliconangle.com/blog/2014/04/01/ibm-cisco-
ge-att-form-industrial-Internet-consortium/
(Wheatley, 2014)

14. Brech, B., Jamison, J., Shao, L., & Wightwick, G.
(2013).   The   interconnecting   of   everything.
Retrieved   March   30,   2015   from
Http://www.redbooks.ibm.com/redpapers/pdfs/red
p4975.pdf

15. Gartner (2014). Gartner says a thirty-fold increase
in Internet-connected physical devices by 2020 Will
significantly alter how the supply chain operates.
Retricved   March   25,   2015   from
http://www.gartner.com/newsroom/id/2688717

16. Robinson, J. (2014). For GE CEO Jeffrey Immelt, the
Internet of Things is about saving the planet.
Retrieved   April   8,   2015,   from
http://pando.com/2014/05/15/for-ge-ceo-jeffrey-
immelt-the-Internet-of-things-is-about-saving-the-
planet/

**Chapter 10: Ones and Zeroes –
Everything Digital**

To speak in generalities, most people perform or participate in a daily routine. If a poll were taken asking people how Big Data impact their everyday lives, one would gamble to say they would not know how to respond. How "Big Data" are used in our daily lives and the impacts it has on the average person will be explored further in this chapter. The world as we know it is generating data on an unprecedented scale. As the world becomes ever more interconnected through the use of technology, innovative companies are starting to realize and exploit the untapped potential and value of Big Data.[17]

Not only is the immediate or "real-time" value being utilized, the significance of archived data is being harnessed in new ways as well. "In the Big Data age, data are like a magical diamond mine that keeps on giving long after their principal value has been tapped." [17]

Background

Before "Big Data" became a term that filled the technology publications and scholarly journals, Ponschock, (2007) researched the phenomenon of digital storage of personal data. His research examined the storage of the digital transactions and the direction or directions that digitally encoded personal data flow into and out of containment vaults known as digital storage.[20] "Internet companies, including Google, routinely turn over authentically private information in response to focused warrants and subpoenas from prosecutors and litigants."[14] "The escalating information society, together with the increasing reliance on technology for intelligence gathering and surveillance...has resulted in a growing public policy debate regarding the balance that should be forged between technology and privacy."[18]

This chapter illuminates the life tasks that take place in the technology domain which, by their execution, create digital history that is stored in a multitude of locations such as digital vaults.[20] Technological conveniences provide an opportunity for the provider to routinely collect personal information generated by daily transactions. Collection begins at the point a consumer signs an application and continues through a continuum of infinite usage. Transactions collected and stored in digital "vaults" approach a permanent life span. It can be argued that a combination of the two constructs, collection and longevity, results in greater volume of data in digital form than was present in tangible forms of the past (Figure 30).[20]

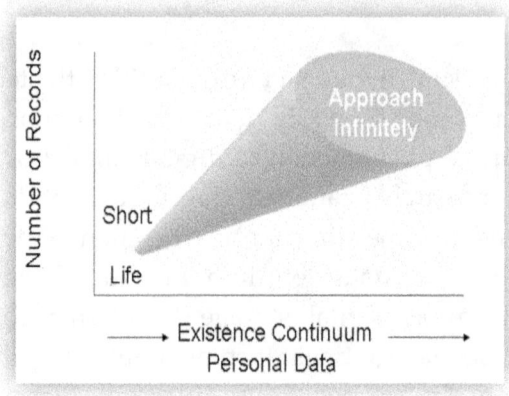

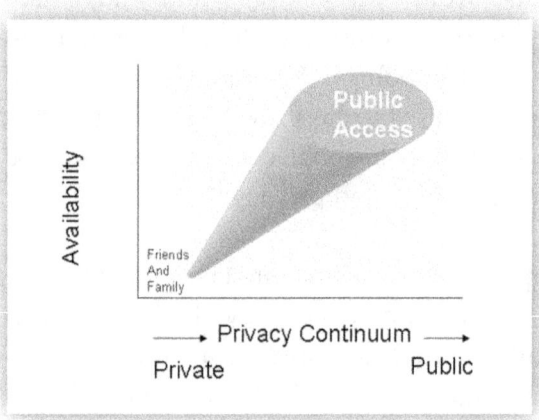

Figure 30: Personal Data Permanency
© 2007 Ponschock[20]

More importantly, the Internet has broadened access to collected, aggregated, and mined personal information. At this point in the chain of logic, there is no distinction between data collected through the daily transactional activity, e-mail, or conversations via the many social networking facilities. Based on the participants' view, it appears that access to personal information is more easily available than it was before digital collection (Figure 31).

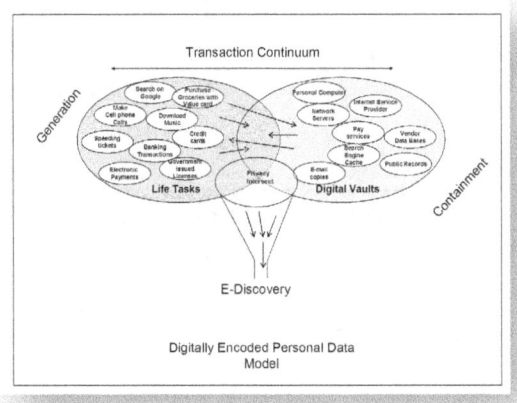

Figure 31: Digitally Encoded Personal Data Model
© 2007 Ponschock[20]

According to SAS.com, "Big Data" is a popular term used to describe growth and availability of data, both structured and unstructured.[2] The amount of data being generated today is truly staggering. With the proliferation of smart phones, tablets, and personal computers, we as a global society are generating enormous amounts of data on a daily basis. Ubiquitous computers generate data on a continual basis. The following table is a sampling.

Auto Computer	Airline Jet Engines	Smart phones	Warehouse conveyors
Airline "Black boxes"	Computer Cookies	Social Media activity	Public surveillance cameras
Private surveillance cameras	Gas meters	Electric meters	Thermostats
Home security system	Company security card	Tax returns	Amazon inquiries
Kindle Reading habits	Google and Bing searches	Credit cards Buying habits	Retail Loyalty cards
Smart Kitchen Appliances	NetFlix	GPS	Water meters

Every minute of every day, data are being generated in some form or another. In one single minute, 48 hours of video is being uploaded to YouTube. Google will receive over 2,000,000 search queries. Facebook users will share 684,478 images on their pages. Apple receives about 47,000 app downloads. And in the minute it has taken to type this paragraph, 204,166,667 e-mails were sent all over the globe.[16]

In 2012, every day 2.5 quintillion bytes of data (1 followed by 18 zeroes) are created, with 90% of the world's data created in the last two years alone. As a society, we're producing and capturing more data each day than was seen by everyone since the beginning of the earth.[3]

With the volume of data being generated rising exponentially, companies are now using this information for a number of purposes. For example, Netflix, the popular online streaming movie service analyses user and usage data to suggest movies an individual might want to watch next, based on movies they have previously viewed. By suggesting movies, the end user would possibly spend more time on the service. It is plausible they would continue to pay the monthly subscription fee as well. Netflix consistently strives to seek continued profits for its shareholders while providing a high-quality service for its subscribers.

In the very near future, self-driving cars could become a reality. Last year, Google CEO Sergey Brin said self-driving cars will be a reality for "ordinary people" in less than five years. With that stated, these cars would possibly generate 1GB of data per second, which on average translates to about 2 petabytes of data per year. Many manufacturers would be able to analyze and interpret the data being produced. This would allow great opportunities to spot mechanical problems before they happen and even schedule repairs.[22]

Story Time

How does "Big Data" impact daily lives? Once upon a time there was a human named Bob. As noted previously, we all have set routines beginning with getting ready for work or play. As an example of how Big Data plays a part in our daily lives, the examination of the routine of a hypothetical character named "Bob" will be explored.

Bob wakes up at 0600 Monday through Friday to make his daily commute to his Network Engineer job. When he rolls over to stop the alarm on his iPad, he quickly views the daily weather provided by the Weather Channel application. Behind the scenes, the application is presenting information based on large-scale, algorithmic-type modeling. This information is stored across 13 data centers. Users such as Bob place a demand of up to billions of computer-based data requests per day on the system. The Weather Channel relies on accurate data analysis to provide its customers with highly detailed and precise weather information. All of this takes place within as little as 10 milliseconds of latency. All so Bob can make an informed decision whether or not to pack an umbrella for the day.

The daily adventure continues as Bob has his breakfast and proceeds out of the door to make his daily commute to work. Bob lives 28 miles from his work, and it will take him 42 minutes to arrive at his destination. The commute is normally a shorter one but on this particular day, the GPS system Bob uses has warned him about possible traffic delays and has chosen another route for his commute. Bob uses the Magellan SmartGPS ®2930.

The SmartGPS 5390 helps users save time and money by automatically delivering the locations of the lowest gas prices and other real-time valuable local information such as reviews and tips from Yelp and Foursquare, as well as

PhantomALERT red-light camera alerts, weather reports, traffic conditions, and much more.[15]

Companies such a Magellan provide information in real-time to the GPS end user by connecting them with access to other big data collectors and compilers such as Yelp and Foursquare. With a quick search, Bob can query Yelp's databases to find the closest gas station in relation to where he is on his route. Bob has unprecedented access to information stored in numerous companies' enormous databases.

Bob arrives at work and continues with his daily routine. While configuring a Cisco router, he encounters a few issues he is unfamiliar with. With a quick Google search, Bob finds the information he is looking for and is able to complete his task. Google servers and its search engine handles on average, 5,922,000,000 searches per day.[6] To date, Google has indexed over 24 billion webpages.[10] With little thought, he has accessed the planet's largest repository of big data.

As the day draws down and the workday comes to a close, Bob proceeds to head home. Upon arriving, Bob turns on his living room light. He gives little thought to how his electric company is monitoring his use of electricity. His smart meter gathers or collects information and sends updates to the electric company every 15 minutes. Over a period of a year, the electric company compiles all the data generated and collected by Bob's smart meter. The company adjusts his monthly bill based on an average of his use over 12 months.

By analyzing this information and the data collected from its other customers, the electric company can better prepare for peak seasons when there is a high demand for energy. This typically takes place in the summer and winter months. When Bob receives his new adjusted bill, he can access his account online to view his monthly or yearly usage rates. Having this visual representation based on thousands of hours of collected data, Bob can plan to use less electricity during peak hours or peak months. By doing so, he can reduce his bill for the coming year.[23] As the day concludes, Bob turns in for a good night's sleep. While lying in bed, he browses YouTube videos to find easy listening music to help him fall asleep. He has just connected to a site where more than 1 billion users visit the site monthly. Over 100 hours of video content is uploaded every single minute. Bob has access to endless amounts of data in the form of streaming video. YouTube provides its service for free to the end user. The company can monetize its video library through advertising partnerships. External advertising companies purchase ad space, which is displayed at the beginning or throughout the video.

Through complex algorithms and detailed analysis of data, YouTube can pair advertisers with potential customers based on the videos everyone viewed. If Bob clicks on a Jazz video, for example, an advertisement for music instruments might be displayed at the beginning of the video. This is a very powerful tool for advertising agencies. Big Data is instrumental in YouTube's success as well as that of the advertisers. Bob falls asleep listening to his favorite "Blues" artists.[25]

Optional Value

Companies historically have not harnessed the potential of the data in which they collect. Initially, data are collected, analyzed or used, and stored. A shift in this usage pattern is taking place in the business world. Companies with vast data archives are extracting information and repurposing it from its primary use;[17] this is considered to be data's optional value.

During the 1990s, hospitals began to transition from paper based medical records to a digital format known as the electronic medical record (EMR) or the electronic health record (EHR).[19] The enormous amount of data collected has helped healthcare professionals streamline record collections and significantly reduce the warehousing of physical records. Patients are able to transfer from one doctor's office to another without having to maintain a physical copy of their records. This is convenient for the patient and there are significant savings in the form management, storage, paper products, etc. for both the doctors and patients.

The optional value gained from converting physical records to digital records is being realized according to journalist Kelly Kennedy of USA Today. Secondary information utilized by insurers will soon reassess how they predict costs. Patients will let doctors know what medications won't work with their particular genomes, and researchers will look at hospital records in real time to determine the cheapest, most effective ways to treat patients. The analysis of large sets of data, such as medication usage or hospital readmissions, has enabled health care providers and policymakers to make smarter decisions and predict future trends.[12]

Another example of the secondary information bringing value to the health care industry is evident with companies such as Kaiser Permanente. The company created a new computer system named Health Connect. The system is a data exchange across all medical facilities promoting the use of EMR's. The integrated system improved outcomes in cardiovascular disease and achieved an estimated $1 billion in savings from reduced office visits and lab tests.[1]

Running on Fumes "Data Exhaust"

Companies have found value in the reuse of data. This can be seen particularly in the data collection and reuse of data by grocery stores. Fresh & Easy, a grocery store chain located in the western United States offers a "friend's rewards" card to shoppers. Each time a person swipes the card, the company collects information specific to the shopper.

All the items purchased are tracked. The company can analyze the initial data to learn more about shopper's trends. Which items are selling and which items are not? The same information is reused and supplied to product producers who in turn are able to provide coupons specific to the customer. These coupons are printed at the register. Fresh & Easy in partnership with its suppliers are able to reuse the data to target market each shopper with the hopes of increased sales.[13]

Not Just Fumes

Data exhaust is unstructured information or data that is a by-product of the online activities of Internet users; [5] it refers to data that is shed as a byproduct of peoples' actions and movements in the world[17]. The power of data exhausts can be witnessed by the creation of Bounce.io. Scott Brown, founder of Bounce.io was able to harvest 20-30 million bounced e-mails a day and monetize what essentially are data exhaust. According to TechCrunch.com, Bounce.io takes bounced e-mails and does two things. It adds advertising to the bounced e-mails that actually came from people who, for whatever reason, were sending an e-mail to the parked domain. And the e-mail from spam bots gets sold to data security companies that view it as a fresh source never tapped before. For them, it's an organic Big Data honey pot that is continually refreshed. The story here is about a company that stands to make a fortune in e-mail advertising. But even more so, Brown's experience points to the power of "data exhaust," all that excess data that trails us. It is data that is one-dimensional by itself. In its mass, the data has economic meaning and reflects more about us than we might think.[29]

Depreciating Value

Although data can be used time and time again, in some businesses, the information being used may no longer be viable. Data, in essence, can have a "shelf life" at times. Companies such as Netflix or Hulu use archived data to offer movie suggestions to their users based on past viewing. If the information is outdated or not regularly screened to authenticate its usefulness, the suggestion algorithm these companies use would be useless or immaterial. Amazon.com, for example, uses sophisticated models to help separate useful data from irrelevant data. This way they are able to determine the usefulness of the data and develop accurate depreciation rates.[17]

Open Data

Governments are some of the largest collectors of data, but they have often been ineffective in using the collected information in purposeful ways. This has led to the idea of "open government" where information collected by the government is openly shared with the public. Open data is considered public data in which people can access to make data-driven decisions.[8] Not all information is shared publically due to privacy concerns.[17] As reported by Joel Gurin of The Guardian all definitions of open data include two basic features: the data must be publicly available for anyone to use, and it must be licensed in a way that allows for its reuse. Open data should also be relatively easy to use, although there are gradations of "openness". And there's general agreement that open data should be available free of charge or at minimal cost. Data from local governments, for example, can help citizens participate in local budgeting, choose healthcare, analyze the quality of local services, or build apps that help people navigate public transport.[8]

Chapter Wrap-up

The potential and value of Big Data are just starting to be realized. Companies are making a fundamental shift from storing data after its initial use, to finding creative and useful ways to harness its true value long after the data are collected. By recognizing data's option value, the health care industry is making strides to mine information for patient medical records. Information is being shared with insurance and pharmaceutical companies to help elevate patient care, increase profits, and streamline processes. The concept of open data has proved beneficial for a number of companies. For instance Opower works with over 85 energy utilities – including 17 of the 20 largest in the U.S. – to provide millions of people across the country with a better understanding of how they use energy. Opower home energy reports have helped Americans save over 2.6 terawatt hours of energy and more than $200 million on their energy bills. Opower relies on open and transparent U.S government data from a number of sources to create value for its utility partners and millions of their residential customers across the country.[11]

As new methods are developed to extract and utilize archived data, along with "real-time" data, companies will begin to appreciate the true value of big data. Whether we think about it or not, we are using resources provided by Big Data to enable us to carry on with our daily routines. We now see Big Data being integrated in most every facet of our lives. From the collection of information our modern cars are producing, to grocery shopping, and even our homes becoming "smart homes" with data being generated daily. Information is being collected and analyzed by businesses to make better products, predict weather patterns, flu outbreaks, targeted advertising, etc. The uses for Big Data are seemingly endless as companies see the great value in data collection. As the access to digital content becomes more available for the globe, the increase in data output is going to be exponential. As we move forward as a global society, Big Data will become more and more integrated in our daily lives and routines. The continued growth of Big Data will allow companies to explore techniques to interpret, extract, and use stored information in ways never imagined before.

End Note

1. Basel Kayyali, D. K. (2013, April). *The big-data revolution in the US health care: Acceleration value and innovation.* Retrieved from McKinsey: http://www.mckinsey.com/insights/health_systems _and_services/the_big-data_revolution_in_us_health_care

2. *Big Data.* (2014, March 1). Retrieved from SAS: http://www.sas.com/en_us/insights/big-data/what-is-big-data.html

3. Conner, M. (2013, July 18). *Data on Big Data.* Retrieved from Marcia Cornner: http://marciaconner.com/blog/data-on-big-data/

4. Constine, J. (2012, August 12). *How Big is Facebook's Data? 2.5 Billion pieces of content and 500+ terabytes ingested every day.* Retrieved from Tech Crunch: http://techcrunch.com/2012/08/22/how-big-is-facebooks-data-2-5-billion-pieces-of-content-and-500-terabytes-ingested-every-day/

5. *Data Exhaust.* (2014, April 22). Retrieved from Dictionary.com: http://dictionary.reference.com/browse/data+exhaust

6. *Google Stats.* (2013, December 20). Retrieved from Statistic Brain: http://www.statisticbrain.com/google-searches/

7. Greenwald, G. (2013, June 9). Edward Snowden: The whistleblower behind the NSA surveillance revelations. *The Guardian*, p. 1.
8. Gurin, J. (2014, April 14). *Big data and open data: What's what and why does it matter.* Retrieved from The Guardian: http://www.theguardian.com/public-leaders-network/2014/apr/15/big-data-open-data-transform-government
9. Hackett, K. (2013, September). Edward Snowden: The new brand of whistleblower. *Quill*, pp. 26-31.
10. *Internet size.* (2014, March 27). Retrieved from The Size of the Internet: http://www.worldwidewebsize.com/
11. Kalin, I. (2013, July 15). *Who uses open data?* Retrieved from Energy.gov: http://energy.gov/data/articles/who-uses-open-data
12. Kennedy, K. (2013, November 24). *Analysis of huge data sets will reshape helath care.* Retrieved from USA Today: http://www.usatoday.com/story/news/nation/2013/11/24/big-data-health-care/3631211/
13. *Let's stay friends.* (2014, April 23). Retrieved from freshandeasy: http://www.freshandeasy.com/about-us/about-friends/
14. Liptak, A. (2006, January 26). In case about Google's secrets, yours are safe. New York Times Online. Retrieved January 28, 2006, from http://www.nytimes.com/2006/01/26/technology/26privacy.html
15. Magellan. (2014, March 27). *News and Events.* Retrieved from Magellangps:

http://www.magellangps.com/Newsroom/Press-Releases/Press-Release-March-27-2014

16. *Mashable*. (2012, May 1). Retrieved from Mashable: http://mashable.com/2012/06/22/data-created-every-minute/

17. Mayer-Schonberger, V. (2013). Value. In K. Cukier, *Big Data: A revolution that will transform how we live, work, and think* (pp. 98-122). Boston: Houghton Mifflin Harcourt.

18. Nelson, L. (2004). Privacy and technology: Reconsidering a crucial public policy debate in the post-September 11 era. Public Administration Review, 64, 259-269.

19. Pinkerton, K. (2013, December 13). *EMR History*. Retrieved from AexineArticles: http://ezinearticles.com/?History-Of-Electronic-Medical-Records&id=254240

20. Ponschock, R. L. (2007). Computer technology, digital transactions, and legal discovery: A phenomenological study of possible paradoxes (Doctoral dissertation, Capella University, 2007) (UMI No. 3246872).

21. Schow, A. (2014, July 20). NSA now admits Edward Snowden sent more than one e-mail, but won't disclose them. *Washington Examinar*, p. 1.

22. *Self-driving cars*. (2013, July 23). Retrieved from computerworld.com: http://www.computerworld.com/s/article/9240992/Self_driving_cars_could_create_1GB_of_data_a_second?pageNumber=1

23. *Smart meter*. (2013, December 11). Retrieved from APS: http://www.apsmeters.com/faq.php

24. Snowden, E. (2014). *Biography.com*. Retrieved July 15, 2014, from Edward Snowden: www.biography.com/people/edward-snowden-2126289

25. *Statistics*. (2014, March 28). Retrieved from YouTube: http://www.youtube.com/yt/press/statistics.html

26. Timm, T. (2014, June 5). *Four ways Edward Snowden changed the world – and why the fight's not over*. Retrieved July 14, 2014, from Gardian: http://www.theguardian.com/commentisfree/2014/jun/05/what-snowden-revealed-changed-nsa-reform

27. *Whistleblower*. (2014, July 15). Retrieved July 15, 2015, from Investopedia: http://www.investopedia.com/terms/w/whistleblower.asp

28. Whitaker, L. P. (2007). *The Whistleblower Protection Act: An Overview*. Washington DC: Congressional Research Service.

29. Williams, A. (2013, May 26). *The power of data exhaust*. Retrieved from TechCrunch: http://techcrunch.com/2013/05/26/the-power-of-data-exhaust/

Chapter 11: "Digital Identity Crumbs©"

This chapter will examine and evaluate the various aspects associated with digital footprints that leave everlasting remnants that can easily identify the person who has created these imprints. The imprints have been termed "Digital Identity Crumbs ©".

Through exhaustive examination of existent research, current events and emerging trends, the research has concluded that significant exposures exist in the protection of identities, personal data and long lasting ramifications of their use. It is the ongoing intent of follow on research to continue capturing all major aspects related to the digitization of society (a phrase coined by the authors in the initial vestiges of this longitudinal study), and provide conclusions as to the viability of control, use and leverage of such data creation.

Background

Fox News published on February 10, 2014: "The U.S. government reportedly is ordering some drone strikes based on the location of terror suspects' cell phones -- without necessarily confirming the location of the suspects themselves -- raising concerns about missiles hitting unintended targets."[18,50] The research in this paper is not intended to argue or support the appropriateness of these actions but rather to point out that existent digital footprints which the authors term "Digital Identity Crumbs©" are being dropped by each individual, leaving a trail that is easily tracked (and potentially used for nefarious purposes) by others.

This chapter expands on the phenomenological study: *Computer Technology, Digital Transactions, and Legal Discovery: A Phenomenological Study of Possible Paradoxes* [43] that examined the sources of existing digitally encoded personal data and how the innovations projected along a 10-year continuum with a potential impact of creating new challenges for control, use, and leverage of that information and data. The research attempted to better understand how digitally encoded personal data might impact the common person or organization. The model included an examination of how the trail of digital transactions can also be compiled into a personal dossier and even used in litigation, or for purposes other than the original author's intent. In addition to secondary usage, "Digital Identity Crumbs©" are left on the digital trail are "Data Exhaust". "Data Exhaust" is a byproduct of society's actions and movements. "Data Exhaust" is being harvested, recycled and even becomes learning data.

227

Amazon's recommendation capability is a notable example.[51]

In the seminal study, an argument was developed that society has embraced the conveniences and efficiencies that technologies have advanced. The research has shown that the use of the technological conveniences have provided an opportunity for personal information generated by these transactions to be collected. Collection begins at the point a consumer signs an application and continues through a continuum of infinite usage. The research provided clarity that transactions collected and stored in digital format approach an infinite permanency. It can be argued that a combination of the two tenets, collection and longevity, result in greater volume of data in digital form than was present in tangible forms in the past, inferring from the expert participants' opinions and knowledge, a conclusion that there has been an increase in data collection, data permanency, and public access to the personal information that may be readily available via the Internet.[43] The study is now several years old, and the research herein is intended to employ the seminal findings while bringing current content to that research. The nine themes depicted in Figure 32 emerged while analyzing the narrative responses of the subject matter experts in the seminal research.

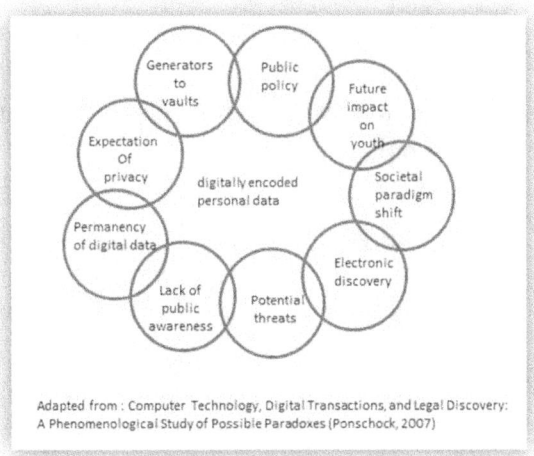

Adapted from : Computer Technology, Digital Transactions, and Legal Discovery:
A Phenomenological Study of Possible Paradoxes (Ponschock, 2007)

Figure 32: Digitally Encoded Personal Data Themes[43]

Digital records are at the very core of society's business environment, and personal lives.[41] Even a small company can easily store terra-bytes of digital data. A terabyte is a measure of computer data storage equivalent to one thousand billion characters of information. An individual leaves a trail of personal data while buying groceries, banking, Facebook, or just driving past video cameras. These "Digital Identity Crumbs©" are contained within an individual's personal computer, throughout the company's network, and beyond. Each day the news has been inundated with corporate litigation that surfaced illegal activity and injurious information found in e-mails, data, and even deleted records, or exposed a breach in securing these records.[47] The "Digital Identity Crumbs©" and the information that they contain can be aggregated, and bought and sold[51] on both legitimate and dark markets.

229

This research examined the sources of existing "Digital Identity Crumbs©" and how the innovations projected along an analytical continuum might impact those sources or create challenges for businesses, individuals, and society as a whole.

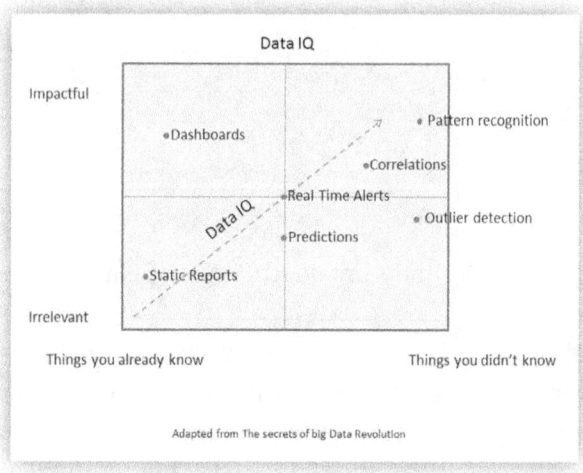

Figure 33: Data IQ Model[51]

The research spawned an understanding of how "Digital Identity Crumbs©" may impact society. The model includes an examination of how the trail of digital transactions can be compiled into a personal portfolio and those portfolios combined with other demographic data points can be used for purposes other than the owners' intent. The secondary uses of these data may not even have been imagined during the initial collection period. [51] This analysis is a continuation of a broader longitudinal and (conceptually qualitative) meta-analysis research related to the ongoing paradigmatic shift[28] based on technological and societal transformations associated with "digitization". This chapter is a data point toward understanding how humankind is on a continuum to a transformation - the *Digiperson*.

Benjamin Franklin with his kite and key may have created a greater impact than that of the Industrial Revolution.[33] Electronic technology the forbearer of the digital era has dramatically influenced society as a whole. Scholars and common citizenry alike have studied with intrigue the obscure force of electricity. Some are bewildered and others are overwhelmed by this phenomenon. Consumers have been using electronic devices for the greater part of their lives, often taking for granted their existence. Electronic devices have crossed the line of tools to ubiquitous computers. These same individuals are not always cognizant of the trail of "Digital Identity Crumbs©" that are left behind and how others can misuse these digital DNA markers.

Back in 2010, Eric Schmidt then CEO of Google – stated that we now create as much data every 2 days as we did from the dawn of man through 2003. (The Big Data Revolution). We create about 2.5 quintillion bytes of data daily. Google receives 2,000,00 searches every minute of every day. As this research will quantify, even these searches leave "Digital Identity Crumbs©" - a personal identity trail.[24] The following table includes some generators of "Digital Identity Crumbs©".

Table 1: Digital Identity Crumbs© Generators

Video Camera – Public	Video Cameras – Private	Fast pass cards	Credit cards with tap chip
Credit card transactions	Financial transactions	Toll booths	Caller ID
E-commerce transactions	Internet Cookies	Internet history	Face book and other social media postings
Social media meta-data ownership	Making financial transactions with cell phones	The computer in your car	Geico's auto "Snapshot"
GPS'	Cell Phone locator chip	Retail loyalty cards	Security checkpoints
Border crossings	License plate readers	E-mail	ATM machines – transactions and cameras
Video cameras and the public crowd sourcing	DNA	Finger prints	Library card
TSA	GOOGLE mapping cars and WIFI	Toll booth EZPASS tags	Legal/Government documents
Driver's license	Marriage	Passport	Student loans
Mortgages	Rental agreements	NSA	Unintended consequences
Security breach	Unwanted advertising	E-discovery	Tweets in the library of congress
Google	Search engines	AMAZON	

Digital footprints and the corresponding volumes of statistics are not just about the other guy. According to the PEW research center, the majority of users know that there is a plethora of personal data about them on the Internet (Figure 34).[45]

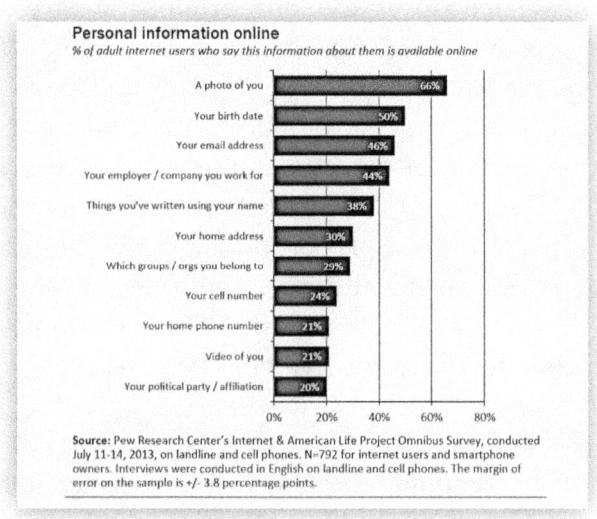

Personal information online
% of adult internet users who say this information about them is available online

Information	%
A photo of you	66%
Your birth date	50%
Your email address	46%
Your employer / company you work for	44%
Things you've written using your name	38%
Your home address	30%
Which groups / orgs you belong to	29%
Your cell number	24%
Your home phone number	21%
Video of you	21%
Your political party / affiliation	20%

Source: Pew Research Center's Internet & American Life Project Omnibus Survey, conducted July 11-14, 2013, on landline and cell phones. N=792 for internet users and smartphone owners. Interviews were conducted in English on landline and cell phones. The margin of error on the sample is +/- 3.8 percentage points.

Figure 34: Personal Information Model[45]

Rights and Principles in a Cyber Era

As technology becomes embedded and approaches invisibility, our daily lives and personal privacy may be pressured or even partially surrendered.[22] The concern over leaking digitally encoded personal data has stirred headlines from many privacy action groups. As individuals browse through Web sites, parties on the other end of those sites may be browsing through their personal information and using that information for their own purposes.[3] The Privacy Rights Clearing House estimated in a posting last updated November 4, 2005, that more than 51 million United States citizens may have had their personal data compromised since February 2005.[47] Visionaries predict that benign computer intelligence may soon be in our homes, cars, and workplaces.[17] Intel and Philips have spent millions of dollars developing the concepts underlying this model. Recent reports of the 2013 Target credit card breach by Bloomberg reported that the Target breach may have affected the equivalent of 1/3 of the population of the U.S. and as for the Ebay breach of 2014 "The magnitude of the reported eBay data breach could be of historic proportions, and my office is part of a group of other attorneys general in the country investigating the matter," stated Florida Attorney General Pam Bondi. These massive breaches exemplify a growing trend line.

As the computer becomes less visible, information appliances will expand in numbers and become embedded in objects used in unambiguous ways.[39] Recently, the Progressive auto insurance company released Information appliances interact in fresh and creative ways.[39] The information appliance also inflicts pressure on the law of transactional volumes. Amaravadi affirmed that the number of digital transactions will increase with the maturation of society. For example, global credit card transactions translate into $213 billion and were projected to hit $393 billion by 2010. Visa transactions totaled 43 billion entries in 2004.[57]

As information appliances meld into our consciousness, their transactions take on forms that are common, frequent, and forgotten. Ubiquitous transaction generators are telephone calls, digital pictures, and e-mail messages. Automobiles commonly inform the service department through wireless connections that the check-engine light has been illuminated. Automobiles produce vast amounts of data whether driven or parked. Acceleration rates, braking distances and battery charging needs for electric cars or hybrids are only a few of the sensory data points being continuously monitored and tracking data collected. These numerous sensors continue to stream data.[38] Through a wireless link to remote diagnostics, the performance of the automobile's critical components can be analyzed.[32] "The data-recorders that are in most cars capture the activities of a vehicle a few seconds prior to an airbag activation have been known to 'testify' against car owners in court in disputes over the events of accidents."[51] Data points are being generated by almost every action of our very existence. Insect drones have also become a reality. A drone with a 16.5-centimeter wingspan can carry a camera, communications systems, and an energy source.[42] A swarm of insect drones can collect "Digital Identity Crumbs©" without us observing their presence. Combined with facial recognition and Wi-Fi interception our every move can be datafied.[51]

As fast as the Internet gains ethical use and acceptance, individuals, companies, and even countries have found ways to exploit its technology. The Chinese government pressured Yahoo to relinquish records, resulting in the arrest and conviction of at least two dissidents.[15] The National Security organization (NSA) a Spy agency of the

U.S. government has intercepted cellphone transmissions of even foreign diplomats. "The NSA snooped on as many as 122 foreign heads of state in 2009, ranging from Merkel to Ukranian Prime Minister Yulia Tymoshenko."[16]

Innovations transition through a present state and mature into new developments that are then positioned to begin the maturation cycle.[39,55] Many of today's electronic or digital innovations are ready to begin the metamorphic transition into levels that were previously imagined only by science fiction writers, visionaries, and futurists.[30,37] Straddling the life cycle of today's technology with the vision of tomorrow provides a balance between the deliverable and the abstract nature of a vision.[39] The telephone provides an example of this principle. Alexander Graham Bell's notebook entry of March 10, 1876, describes his successful experiment with the telephone. Speaking through the instrument to his assistant, Thomas A. Watson, in the next room, Bell utters these famous first words, "Mr. Watson, come here. I want to see you."[5] The telephone has matured from the operator-connected device, to rotary dial, to the touch-tone phone. The cordless phone began to provide freedom of movement denied by the wired telephone because it allowed for mobility.[6] This wireless capability matured into the cell phone of today, which delivers video streams and e-mail. While the cell phone was maturing, the Internet and Web phone also came into focus. Voice over Internet Protocol (VoIP), both on private networks and across the public Internet, has changed the revenues models of many telephone companies.[14] As the cell phone industry grows, the potential privacy violations also increase. Cell phones naturally leave "Digital Identity

Crumbs©". Rep. Phil Montgomery, R-Ashwaubenon is noted as saying "This is a personal infringement on people's privacy."[52] Montgomery also said "It's frightening just how easy it is to obtain someone's phone record. Basically, all anyone needs is a name, number, and address for the number they want to check. With that information, all that's required is a little money and access to the Internet."[36]

"The NSA collects nearly 5 billion records a day on the locations of cell phones overseas to create a huge database that stores information from hundreds of millions of devices, including those belonging to some Americans abroad."[19]

Beyond Data's Original Use

The chart below shows how the 55% of Internet users who have taken steps to hide from someone or an organization compare with those who have not tried any avoidance strategies when it comes to key pieces of personal information that are available online. The Internet users who try to avoid others also often have the most personal information available online.

For years now the citizenry has struggled with the issue of who owns data. The courts with their recent ruling are well on their way to establishing an answer. Facebook tracks a massive amount of data from each of their users. In 2012 it was reported that Facebook collects over 500 terabytes of data every single day.[9] The data Facebook record includes 2.7 billion "Likes," and 300 million photo uploads a day. In addition, Facebook also scans about 105 terabytes a day every thirty minutes.

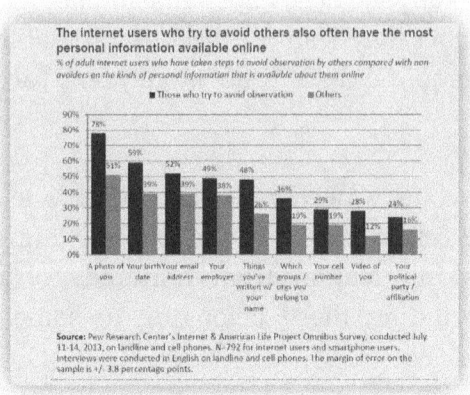

Figure 35: Data Avoidance Research[45]

Data are typically gathered with a specific purpose in mind.[51] The significance produced from using the data as intended will produce primary value. Extensible data, or data preserved in a way others can easily access and use, is potentially important to unlocking its secondary value. Value from "Digital Identity Crumbs©" generators surface in three ways in addition to the primary intended use. First is the basic reuse, where the same data is used to provide insight that was not previously noticed or sought. The second comes from merging of datasets. One set may not have been useful on its own, but when analyzed with another, new insights emerge. The third is finding the recombinant data, which is when two or more unrelated data sets are combined to generate previously dormant value.[51] The typical Internet day is made up of favorite website mines with "Digital Identity Crumbs©" that the

238

private citizen unknowingly leaves behind. In a Wall Street Journal article of 2014 it was predicted the typical online Internet user's activity is tracked over 2000 a day with every click, hover, logon and even delete. Twitter tracks the activity of visitors to over 868 websites out of 2510 of the most popular sites in the U.S., while Facebook tracks the activities on 1205 of those same sites. The secondary uses of these "Digital Identity Crumbs©", individual data, bring enormous value to these organizations.[12]

Permanency

When you swipe your credit card you are making an entry in a data vault somewhere. Ed Gibson, during a speech at the Counter Terror Expo, stated "data that is posted on the Internet should be regarded as permanent after 20 minutes, even if the originator has deleted the file."[48] This is also true with the personal information that you post on social media sites, e-mail, snap chats, and all of the new media Internet communication venues. These digital imprints last forever and can haunt you in ways that you cannot imagine at the point of entry.

E-Discovery

E-mail to some attorneys is equivalent to the proverbial ambulance chaser and analogous to the slippery banana peel.[13] In today's business climate and litigation-prevalent environment, the size of a company no longer dictates the risk of e-mail showing up in a court room. Instant messaging is as vulnerable as e-mail. Whether running a single-person company or a multinational organization, allowing employees to operate e-mail or instant messaging without monitoring or controlling its content potentially places a company's assets, reputation, and future at risk. E-mail and instant messaging are the electronic equivalent of DNA evidence[13] recorded as digitally encoded personal data.[43] The use of pre-trial discovery and particularly e-discovery is growing geometrically. The citizenry is generating digitally encoded personal data at an inconceivable volume. Organizations and most citizenry are unaware that their e-mails or other e-communications might end up being subpoenaed. Similarly, most individuals are unaware that their computer, their network, and even the computer servers that the e-mail is contained on can become involved as evidence in litigation.[60]

What is now known as Zubulake V has become a seminal, precedent setting, and defining e-discovery case. Electronic discovery (or *e-discovery* or *eDiscovery*) refers to discovery in civil litigation or government investigations which deals with the exchange of information in electronic format (often referred to as electronically stored information or ESI).[59] What can't be seen can harm individuals. E-discovery forensic specialists are asked to retrieve deleted correspondence in their attempt to reconstruct a data evidence chain. E-mails are harder to permanently erase than other data files because they often reside in many locations along a computer network.[25] E-mail messages can invisibly reside on an employer's central server, the employee's offline work drive, and other cache files along the delivery path.

The Zubulake landmark decision has taken on new legs with the advancement of Big Data and predictive analytics.

In the case of Bridgestone Americas, Inc. v. Int. Bus. Machs. Corp., No. 3:13-1196 (M.D. Tenn. July 22, 2014) the court approved the request from the plaintiff to use predictive analytics in the review of over 2,000,000 documents. Big Data and predictive analytics will be discussed more in future research.[20]

Headlines From Trend lines

Avid television watchers and media historians of the 1950s may recall "Dragnet" or "Badge 714". During these series sergeant Joe Friday frequently worked the bunco squad. The bunco squad was a special investigative group within the police force that investigated confidence scams. During the roaring 20s and prohibition, gambling parlors sprung up. Thus the detectives that raided the illegal gambling establishments were known as the "Bunco" squad. Bunco a combination of cards and dices was introduced in the United States in 1855 in San Francisco by a shady gambler. Fraud, confidence games, scams, and rip-offs are not new. The confidence game is one of the oldest scams. In the new millennium, these "artists" employ the use of computers over the Internet and the landscape of cyberspace. It is probably safe to say that anyone with an e-mail address has received a notice from what appeared to be a familiar lender, a respectable e-commerce web site, or some other previously used supplier that carries with it trust and security. Phishing is today's "Bunco". It is a cyber-technique used by frauds and thieves that lull the e-mail recipient into believing that one of these trusted partners actually sent the message and actually need updated information from you.

241

Phenomenological studies synergize data collected from subject matter experts or environmental trends on the subject with the end goal of painting clearly illustrated picture of the portents. To support the authors' contention of "Digital Identity Crumbs©" it would be negligent to ignore the headlines of the analyzed period. Headlines can be interpreted to sensationalize an event. However a collection of headlines can also depict a trend line.

Fox news reported on an article originally printed in the Washington Post that the NSA collected 444,743 e-mail address books from Yahoo, 105,068 from Hotmail, 82,857 from Facebook, 33,697 from Gmail and 22,881 from other providers back in 2012. This same report discovered that contact lists permit NSA specialists to create a detailed scatter diagram of a person's life and social connections including political, professional, and religious affiliations.[19] The NSA collects nearly 5 billion records a day on the locations of cell phones overseas to create a huge database that stores information from hundreds of millions of devices, including those belonging to some Americans abroad.[21]

One of the nation's largest information services has begun warning more than 100,000 people across the country they may be objects of a con and subsequent fraud. ChoicePoint Incorporated inadvertently sold personal and financial records to fraud artists apparently involved in a massive identity theft scheme. This security breach resulted in least 800 cases of identity theft.[40] The breach was as recent as 2005, however, in 2014 Target warned approximately 110 million individuals that their credit card data may have been compromised. Shortly after the tsunami arrived, eBay announced a breach that one headline reported as "The 'Inexcusable' Impact on 233 Million Customers." The story went on to state: "exposed customer names, e-mail addresses, physical addresses, phone numbers, and birthdays -- all of which had not been encrypted. Financial information, which had been encrypted on PayPal, was not affected. eBay suggested that all users change their passwords.[46] Exposed personal data can lead to identity theft.

Much is well known about Eric Snowden and his breach of National Security information while working for the US Government. Now a fugitive in Russia, many issues related to information privacy, security considerations and overall disposition of digital crumbs is under scrutiny. Key ramifications include how such vital and secure information could be compromised, as well as the ultimate culpability and disposition for the offender. How safe is digital data, and what can be anticipated for the use of personal data?[58]

Jennifer Barrett reported in a January Newsweek article; *"The e-mail address seems legitimate. The logo and the return address match your bank's, and the official looking letter below warns that fraudulent activity has been detected. ..."*.[4] The unobservant recipient would be tempted to reply to this common "phishing" attempted argued Barrett. The Anti-Phishing Working Group (APWG) is an industry association focused on eliminating the identity theft and fraud describes this "Bunco" as an attack using "both social engineering and technical subterfuge" to steal consumers' personal identity data and financial account credentials. Social-engineering schemes use 'spoofed' e-mails to lead consumers to counterfeit websites designed to trick recipients into divulging financial data such as credit card numbers, account usernames, passwords and social security numbers. Hijacking brand names of banks, e-retailers and credit card companies, phishers often convince recipients to respond. Technical subterfuge schemes plant crimeware onto PCs to steal credentials directly, often using Trojan keylogger spyware. Pharming crimeware misdirects users to fraudulent sites or proxy servers, typically through DNS hijacking or poisoning.[54]

Kevin Mitnick, the computer hacker extraordinaire and felon described in his book "The Art of Deception" that the confidence game is a social engineering skill and is all about gaining someone's trust by lying to them and then abusing that trust for fun or profit. The fraudsters that go phishing need two components, a hook and a catch to reel in.[35] They use SPAM to push out massive volumes of e-mail. SPAM is a broadcast or bulk mailing of e-mail to thousands (or millions) of recipients simultaneously. Perpetrators of such spam ("spammers") gather addresses from web pages, databases, or simply guess by using common domains. SPAM occurs without the permission of the recipients and in the United States spamming is regarded as a crime.[8] Having located this potential catch, "school of phish", the hook needs to be baited in a way that a bite is taken. Phishers typically create a story that makes the potential victim feel obligated to respond. Stories most frequently make the reader believe that an urgent response is needed to one of their accounts from being frozen, or some other unattractive event will occur if this message is ignored. The message normally is worded to build trust, importance, and urgency. The response is designed to be simple and usually asked for the "phish" to simply verify what should already be known by the trusted sender. If the "phish" responds, the reply goes to a web site that will look very much like the one the "phish" believes he is responding to, but actually is redirected to the web site of the fraudster. Fraudsters have not stopped with simple phishing. The bar has been raised to a new level called pharming. Pharmers redirect as many users as possible from the legitimate commercial websites they'd intended to visit and lead them to malicious ones. Michelle Delio submits "The most alarming pharming threat is DNS poisoning, which can cause a large group of users to be herded to bogus sites. DNS -- the domain name system -- translates web and e-mail addresses into numerical strings, acting as a sort of telephone directory for the Internet. If a DNS directory is "poisoned" -- altered to contain false information regarding which web address is associated with what numeric string -- users can be silently shuttled to a bogus website even if they type in the correct URL."[10]

Some information is just given away. Unsubscribing to e-mail SPAM lets the spammer know that there is a legitimate recipient at the address. The spammer will either further exploit that new intelligence or sell the information. Livingston submitted that some offenders are large prominent corporations.[31] He gave Gevalia Kaffe, a subsidiary of Kraft food, as an example.[31] Phishing is a scam that relies on the unsuspecting recipient. In one case, the hoax redirected legitimate inquires for www.GovBenefits.gov to their own site to capture personal data.[49] Other phishing approaches ask the recipient to verify bank account numbers or state that the IRS needs additional information. Many willing-to-please individuals are too eager to provide that information. Everyone must protect themselves from giving away "Digital Identity Crumbs©" that can be used to exploit, defraud, harm, and possibly bankrupt the true owner. Err on the side of caution when asked to validate an account number, social security number, driver's license, or other crumb of potential identity.

Information technology has offered astounding global growth over the past few decades (World Information Trade and Service Organization, 1999). Biotechnology has developed from its nascent research beginnings into commercial applications and uses for social benefits. Biotechnology also brings with it skepticism and cultural concerns.[27] Biotechnology also has the potential for adding security and contributing to identity theft prevention. Applying DNA to access systems may assist in combating identity fraud. One system in development is intended for both authentication and identification of the individual.[34] Using a set of DNA chip cards assigned with a specific DNA code (Group ID), along with the individual's identification information, biometrics, password, and other profile information recorded in the chip's memory. The individual using the card must be a perfect match with the DNA on the chip.[34]

Many of the tasks performed daily create digitally encoded personal data and leave a trail of "Digital Identity Crumbs©". Every day, potential exists to give away personal information and individuals might not even realize when it happens.[44] These various tasks can be generically referenced as digital transaction generators. A growing number of people on the technology side of the digital divide use the Internet frequently. It is increasingly noticeable that certain online retailers are getting smarter and smarter. These storefronts appear to know what was purchased last and many other profile characteristics. This feedback is possible because the retailer stored information about you in information nuggets called *"cookies"*.[53] Cookies are messages that the browser and the server exchange with each other. A cookie exchange with a reputable web site like Amazon results in convenience. It is certainly convenient if Amazon reminds the buyer about previous purchases. However, if the retailer requires a user name and password and stores this information in a cookie, an unscrupulous third party can exploit the digitally encoded personal data. Amazon even records clickstream data showing whether a certain page in a digital book was heavily annotated.[51]

Many individuals drive the toll roads of the nation. To save time, and in some instances money, drivers purchase a toll collection device that electronically records and collects funds as the driver uses the freeway. In New Jersey it is called an E-Z Pass. In the process of acquiring an E-Z Pass, several pieces of personal data are collected. The device reduces congestion and may be perfectly safe,[23] particularly because the passes are issued by a government agency. The three digital generators seem perfectly safe and a convenient part of a technology driven life style. Most users of these transactions or others are aware that there were over 130 recorded privacy breaches between February 15, 2005, and February 23, 2006, affecting 53,416,240 individuals. The breaches in the statistic were recorded because they compromised personal information. Congress is considering several bills requiring notification of individuals when a breach occurs. Over 20 states have similar legislation in place. No single industry or business sector can be singled out. These personal information leaks ranged from hotels and banks to government agencies and those areas detailed earlier.[47] A programmer and user of E-Z Pass alerted officials when he was able to access and view account information via the system's e-mail.[23] As the result of a stolen laptop, 1,400 Safeway Value members' information was compromised.[47]

An argument can be developed that society has embraced the conveniences and efficiencies that technologies have provided during the last 10 years. This research has shown that the use of the technological conveniences have provided an opportunity for personal information generated by these transactions to be collected. Collection begins at the point a consumer signs an application and continues through a continuum of infinite usage. This research provided clarity that transactions collected and stored in digital format approach a permanent life span. It can be argued that a combination of the two constructs, collection and longevity, result in greater volume of data in digital form than was present in tangible forms of the past, and that greater amounts of data exist for a more expansive life cycle.

The Internet has broadened access to collected, aggregated, and mined personal information. At this point in the chain of logic, there is no distinction between data collected through the daily transactional activity, e-mail, or conversations via the many social networking facilities.

This story could have been placed under several difference sections of this chapter but it is a great conclusion for the topic of **Headlines and Trend lines.** In November of 2014 the video of a woman being abducted off the street of Philadelphia made every news channel and paper. Directly related to this chapter is how the crime ended. The reporters indicated that this may very well be the first time abduction was actually caught on video. The trail of physical and "Digital Identity Crumbs©" solved the case.

- Abduction caught on video
- Victim dropped her cell phone and glasses to help police identify her
- Social media provided a plethora of tips including make of car later matched on traffic camera footage.
- Victim kicked out window. This help public spot the car.
- ATM card used
- ATM receipt found
- Convenience store surveillance footage reviewed and the suspect was identified and matched to a previous warrant.
- A traffic camera showed the car and the dealership it was purchased at.
- Dealer had installed a GPS because of the abductor's bad credit.
- The dealer activated the GPS
- The car was located by the GPS
- The story ends with the victim not seriously harmed.

Emerging Economic Footprints

Historically, financial transactions have been tracked based upon physical movement of funds (albeit in many cases using technological interventions). This metamorphosis of digitization as related to financial transactions is now evolving further with potential archaeological considerations minus the physical presence of currency. The Bitcoin has emerged on the scene with much fanfare and equally many questions as to its viability, backing and sustainability. "Bitcoin is the digital currency that thrills nerds, inspires libertarians, and incites the passions of economists who debate the value of money made from nothing but ones and zeros."[56] Regardless of its sustainability, this novel creation of currency is creating quite a stir in the financial markets with multiple vendors experimenting with it for payment of goods and services, and its presence and ability to digitally track it has been expanding exponentially.

Digital cottage industries are now forming, growing, and expanding based on the creation of the Bitcoin. Additionally, a new form of crime is being leveraged from traditional hacking of computer resources, as well as being used for illicit purposes. In 2013, Silk Road (a provider of illegal drugs) was tracked down by the FBI and shuttered based on their acceptance and use of Bitcoins.[29] Subsequently Silk Road went back into business and was hacked out of $2.6 million in 2014.[7] The key aspect of this development is that Bitcoin is serving as a digital footprint upon which it can be tracked, and it appears that the hackers of Silk Road are in the sites of authorities. "With no government oversight or central database to track transactions, how do you prevent fraud?"[56]

Another cottage industry developing is the proliferation of accelerated servers and software to crack the code of mysterious puzzles inherent in the mining of the Bitcoins. Bit coin emerged on the market in 2008 as a more elaborate manifestation of digital currency similar to forerunners such as DigiCash and Bit Gold. A key differentiator appears to be the more elaborate and continually evolving algorithms required to continue mining Bitcoins. It appears that this approach is a never-ending amoeba type of puzzle that keeps recreating and uncovering more Bitcoins for circulation. Hence, the need for advanced technological power and speed to continually solve these elaborate equations resulting in mined Bitcoins!

This digital footprint appears to be a story that is still being written. Its viability and sustainability have yet to be truly tested; but, it also has spurred legal and illegal activities and more importantly has emphasized the evolving ability to track one's steps based on their digital footprints. Additional research is required here to longitudinally track this evolving phenomenon, and its ultimate journey (or demise).

Correlation and Predictive Analytics – (from a pile of "crumbs")

Big Data moves data analysis away from mining, summarizing, and reporting transactions to understanding within and among the aggregation of "Digital Identity

Crumbs©". The sum is more valuable that the individual transactions. Big Data's seminal principal is the correlation and statistical relationships between data values.[51] "Big data marks an important step in humankind's quest to quantify and understand the world. A preponderance of things that could never be measured, stored, analyses and shared before becoming *datafied*."[51] Data is a resource that when used does not get consumed or diminish in value. Data can even be used simultaneously in diverse ways. Almost a story told around the campfire illustrates "What Target Knows". First published in the New York Times magazine back in 2012 and now a legend explains how a predictive analytics scientist and Target employee, Pole, created the pregnancy-prediction model. Some twelve months after Target began using Pole's pregnancy-prediction model,

> *... a man walked into a Target outside Minneapolis and demanded to see the manager. He was clutching coupons that had been sent to his daughter, and he was angry, according to an employee who participated in the conversation.*
>
> *"My daughter got this in the mail!" he said. "She's still in high school, and you're sending her coupons for baby clothes and cribs? Are you trying to encourage her to get pregnant?"*
>
> *The manager didn't have any idea what the man was talking about. He looked at the mailer. Sure enough, it was addressed to the*

man's daughter and contained advertisements for maternity clothing, nursery furniture and pictures of smiling infants. The manager apologized and then called a few days later to apologize again.

On the phone, though, the father was somewhat abashed. "I had a talk with my daughter," he said. "It turns out there's been some activities in my house I haven't been completely aware of. She's due in August. I owe you an apology".[11]

This extrapolation was possible because during the prior month, this high school student may have purchased a large container of unscented lotion, an assortment of supplements such as zinc and calcium and a large purse from Target. Using Pole's pregnancy prediction model, Target even projected that the shoppers due date was in five months.[11]

Chapter Wrap-up

The continuum and velocity of change continues to proceed and accelerate regarding the digital footprints that are being emblazoned for future archaeological consideration. Many prior physical remnants from history are being unearthed to this day, while many others have been lost forever. Far fewer data oriented remnants will disappear forever and remain for future discovery, assimilation and explanation of what has transpired during our lifetime. Although the future historical considerations may be value added and positive in nature, the existence of these "Digital Identity Crumbs ©" also provide current security risks, concerns and lack of regulatory guidance for their protection and use. Individuals remain exposed to nefarious use of their "crumbs", and attention/protection is required via the individual themselves. Future considerations based on this phenomenological and longitudinal study include the various considerations for individual protection, use and dissemination of their personal information. Additionally, all aspects of this phenomenological study must be coalesced so as to provide the holistic picture of risks, exposures, proactive individual action plans, and longer term ramifications for institutions, corporate entities and society at-large.

End Notes

1. Amaravadi, C. S. (2004). The laws of information systems. *Journal of Management Research*, 4(3), 130-136.

2. AT&T. (2005a). 1958: *Laser*. Retrieved October 28, 2005, from http://www.att.com/attlabs/reputation/timeline/58laser.html

3. AT&T. (2005b). 2002: *Privacy bird*. Retrieved October 28, 2005, from http://www.att.com/attlabs/reputation/timeline/02privacy.html

4. Barrett, J. (2004, January 26). When Crooks Go 'Phishing'. Newsweek retrieved March 12, 2014 from http://www.highbeam.com/doc/1G1-112417862.html

5. Bell, A. (1876). Bell's experimental notebook, 10 March, 1876. *Alexander Graham Bell Family Papers*, Manuscript Division, Library of Congress. Retrieved March 11, 2006, from http://memory.loc.gov/ammem/bellhtml/bell1.html

6. Bellis, M. (2006). *History of the telephone*. Retrieved April 14, 2006, from http://inventors.about.com/library/inventors/bltelephone.htm#patent

7. Burns, E., (2014). *Known Bitcoin flaw results in $2.6M stolen from new Silk Road site*. Retrieved September 14, 2014 from http://www.redorbit.com/news/technology/1113072057/silk-road-results-in-bitcoin-theft-021414/#OJ0eEc24uFYI7A1J.99

8. Calburn, T., (2005). *Spam cost billions*. Retrieved May 12, 2014 from

http://www.informationweek.com/spam-costs-billions/d/d-id/1030111?

9. Constine, J., (2012). Facebook ingests 500+ terabytes every day. Retrieved October, 10, 2014 from http://bigdata-io.org/facebook-ingests-500-terabytes-every-day.

10. Delio, M., (2005). *Pharming out scams phishing.* Retrieved September 15, 2014 from http://archive.wired.com/techbiz/it/news/2005/03/6 6853?currentPage=all

11. Duhigg, C. (2012). How companies learn your secrets. Treieved July 29, 2014 from http://www.nytimes.com/2012/02/19/magazine/sho pping-habits.html?pagewanted=all&_r=0

12. Dwoskins, E. (2014). Give me back my privacy. Wall Street Journal. Retrieved September 1, 2014 from http://online.wsj.com/news/articles/SB10001424052 702304704504579432823496404570

13. Eckberg, J. (2004, July 24). E-mail: Messages are evidence: Someone may be watching. *Cincinnati Enquirer* [Electronic version]. Retrieved March 13, 2006, from http://www.enquirer.com/editions/2004/07/27/biz_ biz2a.html.

14. Evalueserve. (2005). P2P VoIP impact on Telecom service providers. Retrieved April 13, 2006, http://joung.im.ntu.edu.tw/teaching/distributed_sy stems/2005EMBA/Impactof SkypeonTelecomServiceProviders.pdf

15. Fagarty, K. (2006). Hypocrisy taints U.S. accusations against Google China. *CIO Central* [Electronic version]. Retrieved March 7, 2006, from

http://www.ciocentral.com/
article/hypocrisy+taints+US+accusations+against+c
hina

16. *Fingas, J. (2014). NSA may have spied on 122 foreign leaders. Retrieved September 11, 2014 from http://www.engadget.com/2014/03/29/nsa-may-have-spied-on-122-foreign-leaders/*

17. Fisher, R. (2004, December 10). Hidden agenda. *Engineer* , 16-17. Retrieved October 20, 2005, from http://www.e4engineering.com/Home/Default.aspx

18. Fox. (2014, February 10). U.S. government reportedly ordering drone strikes based on cell phone location. Retrieved September 10, 2014 from Http://www.foxnews.com/politics/2014/02/10/us-government-reportedly-ordering-drone-strikes-based-on-cell-phone-location/

19. Foxnews. (2013). NSA reportedly collects 5 billion cell phone location records a day. Retrieved July 14, 2014 from http://www.foxnews.com/politics/2013/12/05/nsa-reportedly-collects-5-billion-cell-phone-location-records-day/

20. Gates, K., L. (2014). Court allows plaintiff to "switch horses in midstream" and use predictive coding to review documents initially screened with search terms. Retrieved October, 15, 2014 from http://www.ediscoverylaw.com/2014/09/articles/cas e-summaries/court-allows-plaintiff-to-switch-horses-in-midstream-and-use-predictive-coding-to-review-documents-initially-screened-with-search-terms/

21. Gellman, B., & Soltani, A. (2013). NSA tracking cellphone locations worldwide, Snowden documents show. Retrieved August 23, 2014 from http://www.washingtonpost.com/world/national-security/nsa-tracking-cellphone-locations-worldwide-snowden-documents-show/2013/12/04/5492873a-5cf2-11e3-bc56-c6ca94801fac_story.html

22. Ghadar, F., & Spinder, H. (2005, July/August). IT: Ubiquitous force. *Industrial Management* [Electronic version]. Retrieved October 15, 2005, from http://www.allbusiness. com/periodicals/article/511919-1.html

23. Grygo, E. (2000, October 25). *New Jersey Turnpike electronic toll collection system hacked.* Retrieved March 1, 2006, from http://www.infoworld.com/articles/hn.xml/00/10/25 / 00125hnezpass.html

24. Hemnes, T. (2012). The ownership and exploitation of personal identity in the new media age. *Intellectual Property Law,* *12*(1), 2013. Retrieved October, 10, 2014 from http://repository.jmls.edu/ripl/vol12/iss1/1/

25. Hooper, I. D. (2002, January 26). *Enron's electronic clues: Computer scientists seek to recover "deleted" files.* Associated Press. Retrieved November 12, 2005, from http://www.futurist.com/futuristnews/archive/futu re_courts.htm

26. *http://online.wsj.com/news/articles/SB10001424052702 304704504579432823496404570*

27. Krimsky, S. (2005). From Aslonar to industrial biotechnology: Risks reductionism and regulations. *Science as Cultures, 14,* 309-323.

28. Kuhn, T.S. (1996). *Structure of scientific revolutions.* Chicago: University of Chicago Press.

29. Lee, D., (2104, February). *Silk Road: How FBI closed in on suspect Ross Ulbricht.* Retrieved September 14, 2014 from http://www.bbc.com/news/technology-24371894

30. Lindemann, M. (1999). *Megatrends: Then and now.* Retrieved April 10, 2006, from http://www.gsreport.com/articles/art000088.html

31. Livingston, B. (2006, February 7). *Corporations have an unsubscribe problem.* Retrieved September 10, 2014, from http://www.datamation.com/columns/executive_tech/article.php/3583221/Corporations-Have-an-Unsubscribe-Problem.htm

32. Lynch, C. G. (2005, November). Virtual mechanic. *CIO, 19*(4), 22.

33. Manzano, Y. (1999). *Technology's influence in 20th century life and its difference from industrialization's influence in 19th century life.* Retrieved January 3, 2006, from http://ww2.csf.fsu.edu/~manzano/writing/essays/history/technology/html

34. Margolis, J. (2004). *Applied DNA Sciences introduces its DNA-embedded contactless reader system at the International Security Conference West.* Retrieved April 3, 2006, from http://www.secureidnews.com/news/2004/04/04/

35. Mitnick, K. D., & Simon, W. L. (2002). *The art of deception: Controlling the human element of security.* Indianapolis, Ind: Wiley Pub.

36. Montgomery, P. (2006, January 11). *Legislation to create penalties for phone record sales.* Retrieved February, 15, 2006, from press release, Wisconsin State Representative Phil Montgomery (R-Ashwaubenon).
37. Naisbitt, J. (1982). *Megatrends: Ten new directions transforming our lives.* New York: Warner Books.

38. Noor, A. (2013). *Putting big data to work* retrieved July 24, 2014 from https://www.asme.org/wwwasmeorg/media/Resou rceFiles/Network/Media/Mechanical%20Engineerin g%20Magazine/1013BigData.pdf
39. Norman, D. A. (1999). *Invisible computer.* Cambridge, MA: MIT Press.
40. O'Harrow, R. (2005). *No place to hide.* New York, NY: Free Press.
41. Osterman Research. (2003). *Impact of regulations on e-mail archiving requirements* [White paper]. Retrieved October 15, 2005, from http://www.sdmsoftware.co.uk/xdm-new/assets/pdfs/e-mail.pdf
42. Piore, A. (2014). Rise of the insect drone retrieved February 14, 2014 From http://www.popsci.com/article/technology/rise-insect-drones
43. Ponschock, R. L. (2007). *Computer technology, digital transactions, and legal discovery: A phenomenological study of possible paradoxes.* (Doctoral dissertation, Capella University, 2007) (UMI No. 3246872).
44. Privacy Rights Clearing House, (2014) breaches
45. Rainie, L. et. Al. (2013). Anonymity, Privacy, and Security Online. Pew Research Center. Retrieved

September 19, 2014 from http://www.pewInternet.org/files/old-media//Files/Reports/2013/PIP_AnonymityOnline_090513.pdf

46. Reisinger, D., (2014). eBay hacked, requests all users change passwords. Retrieved September 22, 2014 from http://www.cnet.com/news/ebay-hacked-requests-all-users-change-passwords/#!

47. Rosenberg, B., (2014). Chronology of Data Breaches Security Breaches 2006 – Present. *Retrieved September 10, 2014 from http://www.privacyrights.org/data-breach*

48. Ryan (2011). *Data on the Internet is permanent after 20 minutes.* Retrieved September 14, 2014 from http://www.fedcyber.com/2011/04/21/data-on-the-Internet-is-permanent-after-20-minutes/

49. Ryst, S. (2005, March 1). Phishing: Beware the Internal Revenue scam. *Technology News* [Electronic version]. Retrieved March 1, 2006, from http://www.technewsworld.com/ story/47686.html

50. Scahill, J., & Greenwald, G. (2014). The NSA's Secret Role in the U.S. Assassination Program. Retrieved from https://firstlook.org/theintercept/2014/02/10/the-nsas-secret-role/

51. Schonberger, V., & Cukier, K. (2013). *Big data: A revolution that will transform how we live, work, and think. NewYork:* Houghton Mifflin Harcourt Publishing Company.

52. Schuller, K. (2006, January 13). *5 Investigates: Cell phone records for sale.* Retrieved February 26, 2006, from

http://wfrv.com/consumer/local_story_013093607.h
tml

53. Slayton, M. (1996). Risks of cookies. *Lycos.* Retrieved March 7, 2006, from http://www.webmonkey.com/webmonkey/geektal k/96/53/index4a.html

54. Srivastava, T. V. (2007). *Phishing and pharming – The deadly duo. SANS Institute.* Retrieved October 10, 2014 from http://www.sans.org/reading-room/whitepapers/privacy/phishing-pharming-evil-twins-1731

55. Treloar, A. (2005). *Products and processes: How innovation and product life-cycle can help predict the future of electronic scholarly journals.* Retrieved January 5, 2006, from www.bth.se/elpub99/

56. Vance, A., & Stone, B. (2014). *The Bitcoin-mining arms race heats up.* Retrieved March 20, 2014 from http://www.businessweek.com/articles/2014-01-09/bitcoin-mining-chips-gear-computing-groups-competition-heats-up

57. VISA International. (2004). *Company annual report.* Retrieved November 11, 2005, from http://corporate.visa.com

58. Wemple, E. (2014). NBC News' Brian Williams interviews Snowden, Greenwald. *Washington Post May 22, 2014.*

59. Zubulake v. USB Warburg LLC, 55 FRS 3d 622 (SDNY May 13, 2003).

60. Zweig, M. P., & Goldberg, M. J. (2003). *Electronic discovery: A brave new world.* Retrieved October 25, 2004, from http://www.loeb.com/CM/articles/article57.asp

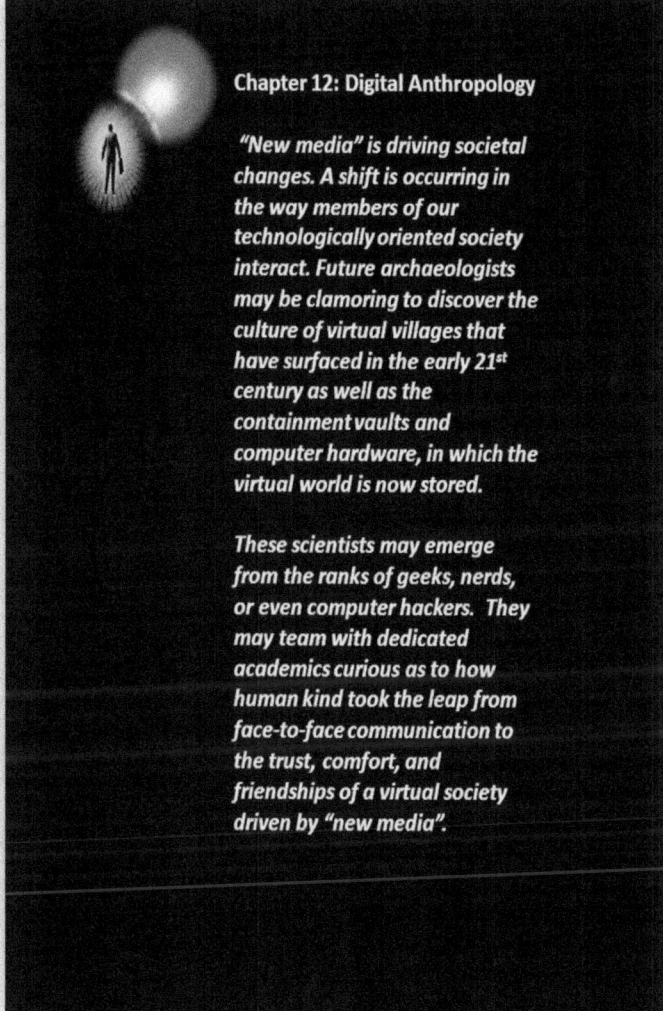

Chapter 12: Digital Anthropology

"New media" is driving societal changes. A shift is occurring in the way members of our technologically oriented society interact. Future archaeologists may be clamoring to discover the culture of virtual villages that have surfaced in the early 21st century as well as the containment vaults and computer hardware, in which the virtual world is now stored.

These scientists may emerge from the ranks of geeks, nerds, or even computer hackers. They may team with dedicated academics curious as to how human kind took the leap from face-to-face communication to the trust, comfort, and friendships of a virtual society driven by "new media".

Background

Technology is changing rapidly, possibly faster than Moore's law[26] ever envisioned. Our society is deluged by this tsunami of change; change is the new constant in our lives. Computer technologies are proliferating computer-mediated communications faster than many can comprehend.[9] As in other periods in history, the nineteenth century technological revolution was founded on prior discoveries in matter and energy associated with earlier innovations.[12]   Comprehending how technology has and will impact the 21st century and beyond, this chapter examines social phenomena through the lens of future archeologists and anthropologists.

Unintended Consequence

The benefits of the electronic "luxuries" that developed countries take for granted do not come without resulting problems. As electronic devices become obsolete, fail to work, get lost, or otherwise become disposed of, they become an environmental nightmare.   Due to these discarded electronics - from TVs to cellphones - landfills and even countries are confronted with the proliferation of e-waste. [32]

> ... a sizeable portion of the e-waste generated in the developed world is exported to developing countries where it is recycled or dumped without any concern for the gross pollution that is being caused. It can be said that if the situation, vis-à-vis e-waste is posing a challenge in most

developed countries, it is alarmingly bad in the developing world. In an attempt to contain the e-waste problem most of the developed world and several countries in the developing world have enacted legislation to curb illegal trafficking and unlicensed recycling of e-waste.[32]

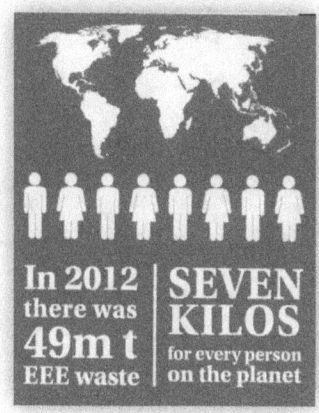

Figure 36: EEE Waste[19]

E-waste in impoverished communities of under developed countries has become an environmental as well as a health problem; the gradual accumulation of heavy metals, such as lead and cadmium in the soils, adversely affects the individuals' underlying health. Electronic e-waste arrives regularly at the ports of Ghana from North America, Europe, and Australia. These broken, obsolete, and unrepairable computers, cell phones, and other e-waste end up being burned in order to extract the precious metals. Often, mercury is used to extract the gold. The unintended consequences are harmful chemicals released into the air, water, and soil.[5]

265

Image 5:

Image 6:

Although shiploads of e-waste end up in developing countries, it is estimated that 75% of electronic items are stored due to uncertainty of how to manage them. These electronic junk piles lie unattended in houses, offices, warehouses, etc., and are normally mixed with household wastes, which are finally disposed of at landfills.[35]

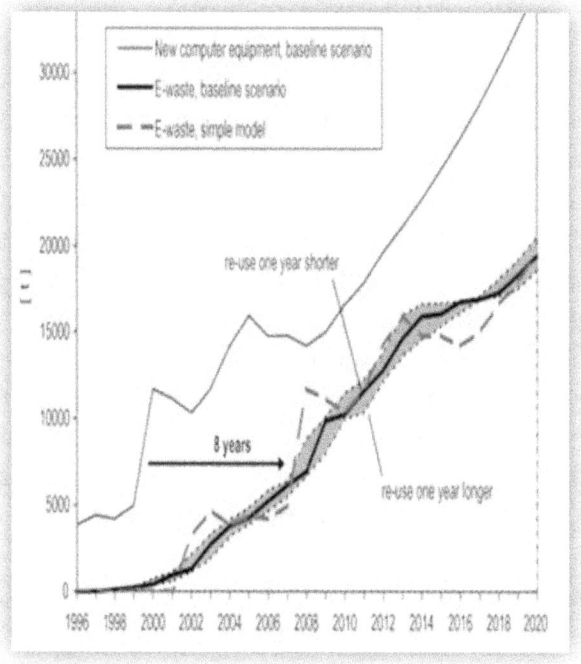

Figure 37: E-waste and Reuse Trends[35]

Legislation by the world community is attempting to get the environmental pains under control. The e-waste that is in the landfills will be a great resource for future archaeologists. Digging the landfills of the world will be one way for the digital prospector to find the futuristic "gold nugget". This chapter focuses on the virtual side of prospecting by future generations - what the future archaeologist will be concerned with, and a little look at anthropology, the study of humans, past and present.

The Archaeologist

Archaeology dates back to the 15th century and in some academic curricula is a sub-discipline of anthropology.[33] Anthropology is seen as the holistic study of who we are as human beings in the physical world amidst cultural diversity, as well as the biological, social, and cultural construct of a period of time, geographic area, or culture.[20] Archaeology is frequently described as the study of the chronology of events and cultural traditions through the recovery, documentation, and analysis of material remains. When thinking about these concepts in regard to exploring virtual digital periods, there is no fundamental difference although the complexity of the documentation and recovery increases multi-fold.[17]

In fundamental terms, archaeology can be illustrated as the recovery of tangible objects, and anthropology, the interpretations, as related to the cultural aspects of the discovery.[21] According to Steibing[34]

> *The archaeologist attempts to deduce facts about bygone societies and events from the physical clues they have left behind. Tools, pottery, houses, temples, art, campfires, roads, and any other remains that show the results of human activity (including such unromantic items as garbage heaps), as well as the skeletal vestiges of humans themselves, all have stories to tell (p. 22).*

The archaeologist interprets the story of people from physical objects. Simply put, "archaeology is the study of mankind's past through the recovery and analysis of its material remains."[34]

To understand life in the 21st century as well as the impact of technology and the "new media", future researchers will require the skills of a computer forensic technician to extract the contents from iPods, iPads, and iPhones (and whatever comes hereafter), each having the capabilities of a mini voice and digital recorder. These are simply one product line of one brand; other digital devices include digital cameras, USB thumb drives, and SD chips. The understanding and potential extraction of what is becoming a virtual society will demand an understanding of computer containment vaults and techniques for removing data from operational as well as non-operational computer systems and storage devices. Anthropologists analyzing the 21st century will use digital mining techniques, computers, software programs and electronic instrumentation, and the potential need to reconstruct the computer operating environments, data formats, etc. required to meticulously recreate, read and interpret the digital information.

Berger and Luckman[4] advanced the theory of social construction when they noted that everyday life is "not only taken for granted as reality by the ordinary members of society in the subjectivity meaningful conduct of their lives, it is a world that originates in their thoughts and actions and is maintained as real." In the late 20th to the early 21st century, individuals have become fast-paced and time-constrained, working longer and harder than they did in prior years.[38] Technology has accelerated the time gap. A new social constructionist shift is occurring in the way members of our technologically oriented society interact.[4,27] It is changing from one of personal relationships between individuals towards technological communication and interaction. Technology has driven the "new media" in recent years and now continues to accelerate. A distinguishing attribute of a true technological reconstruction is that many innovations occur at about the same time.[16] Kuhn explained this shift as a paradigm change.[15] Technology is a complex system in which the actors construct artifacts in a context shaped both by their interests and by the underlying physical nature of their artifacts.[14] The resultant artifacts in this digital world are not objects that will be uncovered by *pick and shovel* years from now; rather, they will be stored digital records of our likes, dislikes and socially constructed realities of this new era. Printed documents of all kinds currently comprise only 0.003% of the total information being generated. [24] Advanced storage devices are rapidly becoming the universal medium for information storage. It is abundantly clear that our habits, friends, and conversations will be digitally recorded and stored for future analysis and scrutiny.

The worldwide web creates an atmosphere where every communication is treated as if it were constructed in a small hometown – where everyone knows everyone else's business. Communities like Facebook, Twitter, the original MySpace, etc. are mere manifestations of this migration to a virtual relationship[29].

Data Mining: *Pick and shovel* not required

A Google, Yahoo or more advanced futuristic search capability will form the basis for "digital archaeology"; it will most likely explain how to perform reconstructive imaging of a "dig" in places like Botai located on the Iman-Burluk River, a tributary of the shim, in Kokshetav, Oblast.[23] Consider that it might be something called..."Digging without the Dirt: Online Excavations". Evolving archeological discipline now includes the use of digital media and digital information to get a clearer image of the society that uses them.[3] Such findings or artifacts from the 21st century could be e-mails, newsgroups and forum postings, databases, digital pictures and/or videos. **Consider** how people several centuries or millennia from now will uncover enough information about us to form a complete picture of what was occurring in the 21st century.[3] With the inception and growth of the digital world, the past is not only being buried under rock and rubble, it is being stored and replicated in digital containment vaults like network servers, personal computers, digital cameras, Personal Data Assistants, and cellular telephones.

In the 21st century, the digital legacy can be replicated and virtually stored in hundreds of locations around the world on multiple media.[29] Less isolated than archeological findings of the past, the emergent archaeologist will search for pieces of information or *cookies* on a server in Bangladesh, a personal computer in Pittsburg, and numerous other globally constructed containment vaults. The current archaeologist pieces together past cultures by interpreting their findings and formulating theories based on the artifacts discovered. The location of the "find" historically has been near the point of origination. For example, Native Americans carved, pecked, chipped, and abraded messages/pictures as etchings called petroglyphs

271

into rocks or resident walls of a cave leaving us clues and insights into their actions and their behaviors.[2] The location of the rock or cave did not change from the time it was created, unless it were transported to another location. In contrast, the discovery of a video in the future may be found a continent away from its origination point.

These emerging archaeologists will hear and see people from the past. Digital containment vaults will have not only the written pictures and words that petroglyphs held; their stories will include videos and e-mails expressing the emotions of their authors and/or readers. While present archaeologists may speculate how buildings were constructed and the tools that were used decades ago, the digital age stores and collects the actual building process for historians to view and analyze. These containment vaults may be part of the aforementioned virtual villages like MySpace, YouTube, iVillage™ Friendster™, Facebook, and Xanga™, or personal blogs on every imaginable subject and discipline, or even on a digital camera in a garage in up-state New York.[31]

Archaeologists analyzing the 21st century digital age will be required to separate factual findings from fantasy, since virtual reality towns or worlds may exist only in cyber space[13]. Since virtual villages or townships are not represented by geography, social class or financial accounting, a new categorization, codification and/or other approaches will be warranted (as well as verification of their veracity). This cyber positioning will be defined by curiosity.[22] The legacy of 100 million subscribers are currently being buried in the form of personal likes, dislikes, dreams, and possibly "dirty laundry" in landfills of virtual villages or virtual communities like MySpace, Facebook, Twitter, etc. Laurie Anderson, a noted musician/artist wrote "Technology is the campfire around which we gather."[18] The 100,000,000 members of MySpace[6] and many similar Internet virtual socialization network communities, interact as individuals do in the physical world leaving artifacts for future discovery while prolonging the legacy and the data trail. In fact, consider examples like the 92-year old housebound individual who wanted to show the world her piano skills through YouTube so it could become a modern time capsule evidencing her prowess[37].

The digital age may also be leaving personal individual information of our human remains through technological advances such as the veri-chip™. The veri-chip is about the size of a grain or two of rice and can be micro-implanted under the skin where it can be then read by a "transponder", a barcode type reader, or tracking device. Although facing privacy barriers, this device could eventually have widespread acceptance, allowing the tracking of an individual's entire medical history and whereabouts for storage and future recovery.[1] Finally, technologies continue to emerge which will further complicate the replication process for future reconstruction of our current civilization. Consider the emerging ethical dilemmas with innovations like 3 parent babies[39]......or, holograms, bionic body parts, and continual technological innovations like non technology[40]....all of these (and more to come) will further complicate the reconstruction of our present day reality.

Browsers

From simple discovery to large-scale research projects, the Internet browser is emerging as one of the greatest mining tools. Although the Internet was introduced before the graphic user interface (GUI) portal, now referred to as the browser, the paradigm shift to a GUI made the Internet more navigable.[15] Access to the Internet created a drastic shift from the way banking, marketing, and advertising had been typically conducted. Some form of browser will undoubtedly provide digital tools for future excavations. Along with the Internet's wide appeal and tremendous utility, personal and private data is left behind as was evidenced in the AOL® privacy scandal where, "21 million search queries also have exposed an innumerable number of life stories ranging from the mundane to the illicit and bizarre"[25]; one of many such examples in recent times.

Search Engines

A browser has limited utility without a powerful retrieval mechanism for the information. Since the early 1990s when the first rather crude program called *Archie* assisted with the retrieval of information from the Internet, data mining has grown into almost instantaneous recovery capabilities from mining search engines like Google, Yahoo, Excite and others (already existent and to come). It is now common for someone to say have you "googled" it when looking for an answer. *Archie*, an early introduction to Internet search capabilities, looked at a list of File Transfer Protocol (FTP) archives created by a basic command in the UNIX operating system providing a searchable database of filenames. *Archie* did not look into the files' contents. In

1991, Mark McCahill and a team at the University of Minnesota advanced *Archie's* earlier introduction. Their new search engine provided a simple way to navigate distributed information resources on the Internet which allowed for enhanced and deeper discovery into digital files by indexing plain text documents. Many of these same text files evolved as websites with the creation of the World Wide Web. Each search engine now uses its own proprietary methodology that presents results to user inquiries. Speed and the number of returned matches now determine the popularity of search engines as well as their long-term viability.

Digging into computer files and networks is frequently referred to as computer forensics. Although the word forensics has been defined as "to bring to court"[36], it is also used to describe the process of retrieving data or information from a computer device. US-CERT, a United States government organization defines the forensic disciple as one that incorporates both constructs of law and computer science to collect and analyze data from computers systems, networks, wireless communications, and storage devices in a way that is admissible as evidence in a court of law.[30] As forensic evidence, the data may have been deleted or the device may have been removed from its original operational unit without compromising it admissibility as evidence.

We submit, in the future, digital artifacts discovered and documented using forensic disciplines and methodologies will be viewed as acceptable in research communities as well. In fact, the meticulous protocols required in forensic digital discovery mirrors that of an archeological "dig", and tomorrow's archaeologist will need to possess expert computer data detection and retrieval skills. The log is the documentation of the "chain of custody" in the archaeologists' discovery.[30] Present day logs used by forensic digital examiners represent step-by-step records of not only their findings, they also include the methodology used in the retrieval. Digital computer artifacts often exist in many formats, with earlier versions still accessible in multiple digital containment vaults (i.e. hard drives, memory chips, etc). Knowing the possibility of their existence, even alternate formats of the same data can be discovered through the scrutiny of a practiced digital forensic excavator.

Future Digital Excavation

Future archaeologists will face many new obstacles. Where the archaeologist of the 20th and 21st centuries had to dig through layers of rock rubble or even garbage, the digital age archaeologist will face different layers of obstacles: computer hardware, data encryption, varied operating systems, password protections, rapidly changing storage types and standards, transported recorded language, abbreviations/codes, etc. The storage of virtual artifacts, while containing great insights of the culture and behaviors of the inhabitants of the 21st century, will include challenges of ensuring the authentication and protection of digital artifacts.

The computer forensic archaeologist will need to follow principles, practices and methodologies that will withstand scrutiny and analysis of others. Three steps will need to be meticulously followed:

1. No alteration or modification the digital artifact;
2. Authentication and log of the recovery;
3. No modification during the analysis[7]

Tools, other than picks and shovels, will be used in these future technological excavations and explorations. Many of these devices may not necessarily exist as of yet. Based on what we currently know, some of the digital tools may be browsers, search engines, and software programs like EnCase™ or ILook™ Investigator. If the artifact is from an UNIX™ based computer, Sleuth Kit™ and HashDig™ may be two of the computer software utilities used to uncover digital substantiation.[7]

There will also be three additional layers of complexity in digital excavations. The first will be password protection. A simple Microsoft Word™ document may be password protected. The author may have also encrypted the document. If the excavator is forensically extracting the data, they may also need to understand machine level byte configurations - the zeroes and ones or American Standard Code for Information Interchange (ASCII) representations. In addition, formats used to record the data are often media dependent varying by operating systems and may be media specific, i.e. diskette versus USB memory stick, which will require expertise to unravel.[30]

Digital Artifacts

The purpose of mining is to uncover something of value. Gold miners looked for the chance of discovery in rich caverns of the precious mineral and "striking it rich".

Archaeologists in turn search for clues that will assist in a better understanding of a prior civilization or culture. The discovered objects are artifacts, objects made by the residents of that era, for example: a tool or ornament, a clay tablet, or even a book. These findings have enormous cultural interest and impact on our understanding of societies and their methods of operation.

Digital miners also seek to uncover artifacts. The technological, digital age will provide the archaeologist and anthropologist in centuries to come two avenues of discovery. The most visible of these artifacts will be tangible objects like cell phones, digital cameras, laptop computers, digital network components, servers, routers, and a myriad of other apparatus. Tangible artifacts will provide researchers tremendous historical value in understanding the conveniences and communication devices used in this era. The digital containment vaults embedded deep within each of these will open vast windows into the culture and civilization of the 21st century. Memory chips in discarded cell phones contain text messages, addresses, and pictures of the past. Computer hard drives contain video, e-mail messages, and memos of mergers and acquisitions. Digital cameras display family back yard parties and world travels. ComputerWorld™ reported that in 2007 there were 500 million stored obsolete computers, which they defined as an archeological "Gold Mine"[28].

Future archaeologists face different issues from those of their predecessors. The location of their find may not have anything to do with the continent on which their discoveries are uncovered. The context in which the message is found may be critical to the understanding of its utility and meaning. There may be many identical copies of the same artifact and understanding their nuances may require finesse because digital artifacts differ in many aspects. An artifact of the past, i.e. a piece of pottery, can only be in one place at one time. A petroglyph can only be written once. The exact message, picture, story, account of an event in the virtual digital world can be copied many times over and surface in a setting that may distort or confuse its original meaning and intent.

The excavation of virtual villages and associated blogs will enlighten the researcher with discussions and opinions. Blogs or web logs are digital debates or commentaries that will bring the past to life for the researcher. Blogs will need to be interpreted with caution; the same vault of artifacts that shows the actual past may represent fantasy or distorted versions of reality. For example, the MySpace accounting of an event can be factual or make believe. The researcher may also find digital information the author believed was destroyed by deleting the entry, e-mail, or document since deleting information from normal resident access does not delete all digital copies or originals that can be uncovered through forensic digital evaluation. E-mail SPAM may also lead the voyager down an incorrect anthropological trail of discovery. The researcher will need to triangulate their findings to determine accuracy based upon historically known facts and other images that may be available to substantiate findings as they learn about the future by excavating and understanding digital records of the past.[8]

Chapter Wrap-up

The virtual villages and towns where cyber dwellers of the 21st century reside will be the excavation sites of the future. Upcoming archaeologists and anthropologists will critique people, communities, and societies through what they discover has been recorded in sound, video, and text as it exists in today's digital society as it is excavated and understood in the future.

End Notes

1. Albrecht K., & McIntyre, L. (2005). *Spychips: How major corporations and government plan to track your every move with RFID*. Nashville, TN: Nelson Current.
2. Austin, D. (2005). *The Stahl site petroglyphs: New observations and comments*. Retrieved June 2, 2007, from http://www.petroglyphs.us/article_stahl_site_petro glyphs.htm
3. Baheyeldin, K. (2004). *Introduction to digital archeology*. Retrieved May, 9, 2007, from http://baheyeldin.com/technology/digital-archeology.html
4. Berger, P.L., & Luckman, T. (1967). *The social construction of reality: A treatise in the sociology of knowledge*. Garden City, NY: Doubleday
5. Bemma, A. (2013, April, 21). *The proliferation of e-waste in Africa (full length documentary)* . Retrieved October 4, 2014 from http://adambemma.com/2013/04/21/the-proliferation-of-e-waste-in-africa-full-length-documentary/
6. Cashmore, P. (2006, August, 9). *MySpace hits 100 million accounts: Mashable social networking 2.0*. Retrieved November 9, 2006, from http://mashable.com/2006/08/09/myspace-hits-100-million-accounts/
7. Davis, C. W. (2004). *Software for efficient file elimination in computer forensics investigations*. (Master Thesis, College of Engineering and Mineral Resources, 2004). (UMI No. 1423966).

8. Denzin, N.K. (1989). *Interpretive interactionism.* Newbury Park, CA: Sage Publications.

9. Dunlop, C., & King, R. (1991). *Computerization and controversy.* San Diego, CA: Academic Press.

10. El-Laithy, S. M. H. (2002). *Literacy practices and language ideologies of an Egyptian American Muslim youth group: An exploratory ethnographic study.* Hofstra University, (UMI 3046287).

11. *Exploring the potteries,* (2006) Retrieved May, 9, 2007, from http://www.exploringthepotteries.org.uk/Nof_web site1/local_history_static_exhibitions/arch_tech/pag es/planning_development.htm

12. Forester, T. (1989). *Computers in the human context.* Cambridge, MA: MIT Press.

13. Garrett, J.J. (2006). *MySpace: Design anarchy that works.* Retrieved June 1,2007 from http://www.businessweek.com/innovate/content/d ec2005/id20051230_570094.htm

14. Giere, R. (1999). *Science without laws.* Chicago, IL: University of Chicago Press.

15. Kuhn, T.S. (1996). *Structure of scientific revolutions.* Chicago: University of Chicago Press.

16. Kranzberg, M. (1989). IT as revolution: The information age. In T. Forester (Ed.). *Computers in the human context* (pp.19-32). Cambridge, MA: MIT Press.

17. Hirst, K.K. (2007). *History of Archaeology – part 1.* Retrieved June 26, 2007 from http://archaeology.about.com/cs/educationalresour/ a/history1.htm

18. Intel Brochure (2004). *Wired for wireless.* Retrieved from

http://download.intel.com/technology/comms/wire less/WfW_overview.pdf

19. Johnson, T. (2014). Prospectors: The next generation. *TCE: The Chemical Engineer* (874), 28-31.

20. Just, P., & Monaghan, J. (2000). *Social and cultural Anthropology: A very short introduction.* Oxford, England: Oxford University Press.

21. LeBlance, S. A., Redman, C. L., & Watson, P. J. (1971). *Explanation in archaeology: An explicitly scientific approach.* New York: Columbia University Press.

22. Luthra, N. (2006). *The "Real" and the "Virtual" in public space.* (Master Thesis, University of New York at Buffalo, 2006) (UMI 1431955).

23. Lynch, R. N. (1974) *Rethinking modernization: Anthropological perspectives.* Westport, Connecticut: Greenwood Press.

24. Lyman, P., & Varian, H. R. (2000). *How much information?* Retrieved August 10, 2002, from http://www.sims.berkeley.edu/how-much-info

25. McCullagh, D. (2006). *AOL's disturbing glimpse into users' lives.* Retrieved June 24, 2007 from http://news.com.com/AOL+offers+glimpse+into+us ers+lives/2100-1030_3-6103098.html.

26. Moore, G. (1965). Cramming more components onto integrated circuits. *Electronics Magazine, 38*(8).

27. Mumford, L. (1970). *Myth of the machine: The pentagon of power.* New York: Harcourt.

28. Pratt, M. K. (2007). *Tech trash: Still stinking up the landscape.* Retrieved June 22, 2007, from http://www.computerworld.com/action/article.do?c ommand=viewArticleBasic&articleId=284601

29. Ponschock, R. L. (2007). *Computer technology, digital transactions, and legal discovery: A phenomenological study of possible paradoxes.* (Doctoral dissertation, Capella University, 2007) (UMI No. 3246872).

30. Potaczala, M. (2001). *Computer forensics.* Retrieved June 10, 2007 from http://chantry.acs.ucf.edu/~mikep/cf/CHS5937-TermPaper.pdf

31. Pickavet, C. (2006) *Balancing the business of MySpace.* Retrieved May 15,2007 from http://www.Internetnews.com/feedback.php/http:/www.Internetnews.com/ec-news/article.php/3582941

32. Premalatha, M. M., Abbasi, T., Abbasi, T., & Abbasi, S. A. (2014). The Generation, impact, and management of e-waste: State of the art. *Critical Reviews In Environmental Science & Technology, 44*(14), 1577-1678. doi:10.1080/10643389.2013.782171

33. Renfrew, C. (1980, July). The great tradition versus the Great Divide: Archaeology as anthropology? *American Journal of Archaeology, 84* (3), 287-298.

34. Stiebing, W. H. (1994). *Uncovering the past: A history of archaeology.* New York: Oxford University Press

35. Tiwari, S. K. & Pandey, V. K. (2013. Removal of arsenic from drinking water by precipitation and absorption or cementation: An environmental perspective. *Recent Research in Science and Technology, 5*(5), 88-91.

36. US-CERT (2005). *Computer forensics.* Retrieved May 15, 2007 from http://www.us-cert.gov/reading_room/forensics.pdf

37. Vascellaro, J. E. (2007). *Grandma has a story to tell ... on YouTube.* Retrieved June 10, 2007 from

http://content.hamptonroads.com/story.cfm?story=
125014&ran=57567

38. Waskow, R. A. (2004) *Free time/free people statement:
Freeing our time.* Retrieved November 14, 2006, from
http://www.shalomctr.org/taxonomy/term/45

39. http://www.bbc.com/news/health-31069173

40. https://www.youtube.com/watch?v=5CqUYBopWL
s

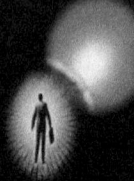

Chapter 13: The DIGIPERSON

We have taken an evolutionary odyssey from Genesis to the archaeologists of the future. This chapter concludes the journey with a deeper discussion on our approach to a speciation event that could transform man into a digital being. This chapter will describe many of the technologies that already exist and some that remain on the laboratory tables and even more that are just in the minds of innovators or possibly even science fiction writers.

The discussion will address Plug-n-Play Neuroprosthetics - implantable devices designed to replace or improve aspects of the central nervous system. Mankind is truly on the edge of a major transformation. For better or worse, the innovations and devices explained have become parts of who we are. Is this the end of one dynasty and the beginning of the speciation dynasty? Each of these dynamic innovations contributes to the transformation - the DIGIPERSON.

Background

From 1974 to 1978, one of the most popular American TV series was called the *Six Million Dollar Man*. The main character, played by actor Lee Majors, was based on a fictional character Steve Austin, a former astronaut who was reconstructed with bionic implants. The *Six Million Dollar* man had super human strength and powers. But more importantly was "alive".

Image 7:

Just a few decades later, the following table lists a few of the "implants" and advances commonly available today. These devices have become part of who we are. Each of these contribute to the transformation - the DIGIPERSON:

Exoskeleton	*Mind-mind interface*
Printed Organs	*Dolly the sheep*
Artificial heart	*Artificial limbs*
Pacemakers	*Speech interface*
Pacemaker	*Bionic Eye*
DNA testing	*Heart valves*

Implantable technology has rapidly progressed. Following is a chronology of implants starting in 1988 with the diaphragm pacemaker.[1]

Topic	Author/year	Article summary
	Implantable neuroprosthesis technology	
	Key publications in implanted neuroprostheses	
Review	Ragnarsson [88], 2008	An overview of the progress made in FES applications, current challenges and suggested future improvements.
	DiMarco [32], 2005	A review of Functional Electrical Stimulation (FES) technology for restoration of respiratory function.
Upper Limb	Peckham [82], 2001	A multicenter cohort trial of an implanted hand neuroprosthesis for persons with tetraplegia with at least 3 years of follow-up. Hand function of 51 adults with C5 or C6 spinal cord injury (SCI) was compared before and after implantation, and with and without the neuroprosthesis. The neuroprosthesis was safe and well accepted, and increased independence.
Lower Limb	Kottink [70], 2007	A randomized controlled trial of an implantable 2-channel peroneal nerve stimulator among 29 stroke chronic stroke survivors. The intervention group received the implantable stimulator for correction of drop foot. The control group used conventional walking devices (ankle-foot orthosis, orthopedic shoes, or no device). FES resulted in a 23% improvement in walking speed, whereas the improvement in the control group was 3%. The study showed a clinically relevant effect of the implantable stimulator on walking speed.
Bladder	Brindley [16], 1994	A review of the first 500 patients implanted with a sacral anterior root stimulators for bladder control. Of 479 survivors, 424 were using their stimulators when last followed up between 3 months and 16.1 years (mean 4 years) after implantation.
	Creasey [27], 2001	A prospective study comparing bladder and bowel control before and at 3, 6, and 12 months after implantation of a neuroprosthesis for stimulating the sacral nerves with posterior sacral rhizotomy among 23 individuals with suprasacral SCI. At 1-year follow-up, 18 of 21 patients could urinate more than 200 mL with the neuroprosthesis; fifteen of 21 had post void volumes of less than 50 mL. The incidence of urinary tract infections, catheter use, reflex incontinence, anticholinergic drug use, and autonomic dysreflexia were substantially reduced at follow-up relative to pre-implantation. At 1-year follow-up, 15 of 17 patients reduced the time spent with bowel management.
Diaphragm	Glenn [41], 1988	A retrospective study of 477 patients who had diaphragm pacemakers implanted for treatment of chronic hypoventilation. A comprehensive analysis of the pacing methods, complications and results from early experiences with diaphragm pacing provided important directions for future applications.

Figure 38: Chronology of Implants[1]

Miniaturizations, nanotechnology, and computer enhancements have accelerated the progress since this research. More subtle changes are being undertaken and now taking place with the human body and brain. The Internet has for most people in the industrialized countries become the conduit for information. The term "google it" has become etched in our lexicon. We are continually being reprogrammed to read and listen in sound bites. Our news, weather, personal messages, and information are delivered in sound bites. Our brain is being altered to

288

retrieve data in short spurts. It is being found that this "google" phenomenon is altering our ability to concentrate on lengthy content. Nicholas Carr, a bestselling writer and researcher in the technology sphere has stated "And what the Net seems to be doing is chipping away my capacity for concentration and contemplation." James Olds, a professor of neuroscience who directs the Krasnow Institute for Advanced Study at George Mason University, says that even the adult mind "is very plastic." Nerve cells routinely break old connections and form new ones. "The brain," according to Olds, "has the ability to reprogram itself on the fly, altering the way it functions."[26] Carr makes a profound statement: "Thanks to our brain's plasticity, the adaptation occurs also at a biological level." [26] The digitization transformation is not only changing the way we exist, communicate, learn, and conduct our daily routines, it is also profoundly rewiring our brains according to Dr. Small and Gigi Vorgan as positioned in their book, iBrain.[27] Carr's, Small's, and Vogan's studies support the theory that the human person is an ongoing speciation transformation at a biological level. Kandle in a paper titled *"The New Science of Mind and the Future of Knowledge"* links neuroscience to the social sciences, ethics, and public policy, free will, delayed gratification, and decision making.[3] The biological changes are being augmented by bionic innovations.[12] We are now in the bionic era.

Bionic Revolution

Approximately a decade ago, scientists/surgeons performed laboratory stage surgery via the Internet. The doctors were stationed in Australia while the surgery was taking place in Southern California.[7] The surgeons used a newly developed Internet-based laser scissor-and-tweezers technology called *RoboLase*.[14,19]

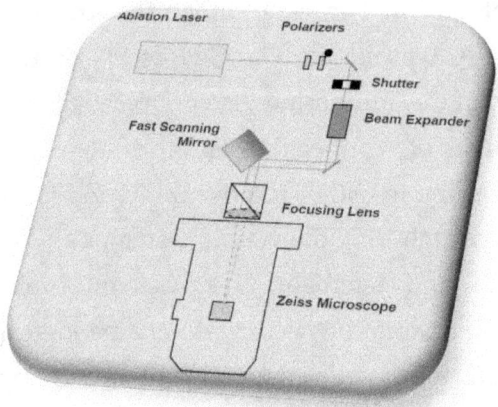

Figure 39: External Laser Light Path of Robotic Laser Scissors Microscope System [14, 30]

Fast forward, and a mere decade later the *da Vinci* Surgical System® is a common medical tool. Surgeons operate through just a few small incisions. The surgical procedures have become so common that they are even advertised on small town television channels. The *da Vinci* System features a magnified 3D high-definition vision system and tiny wristed instruments that bend and rotate far greater than the human wrist. As a result, *da Vinci* enables the surgeon to operate with enhanced vision, precision, dexterity and control. The *da Vinci* Surgical System is a tool that utilizes advanced, robotic technologies to assist the surgeon with the operation. It does not act on its own and its movements are controlled by the surgeon.[4]

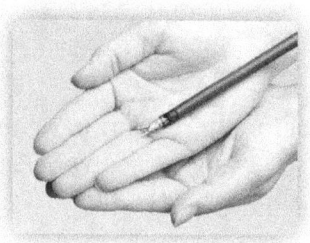

Figure 40: Relative Size of daVinci Surgical System[4]

Breanna Draxler reported "This year marked the turning point for bionic technologies (p. 71)."[5] Taking bionics from tools to human extremities, bionics has become a part of the current conscience. Unique examples include: a dancer returning to the stage after losing a foot and ankle in the Boston Marathon terrorist bombing; a paraplegic completing the opening kick at the World Cup; and, the FDA approval for the first robotic exoskeleton.[5]

3D Printing: "+ and –" Impact on Society

Computer industry publications and many news articles have discussed 3D printing. Articles range from 3D technology being used to produce ceramic guns, plastic ammunition clips, and even titanium parts for jet engines.[34] Although described a bit later, the more important use of this technology for our journey to the DIGIPERSON is in the medical space. Before examining specific uses, it is necessary to define 3D printing at a high level. 3D printing is also referred to as additive manufacturing. 3D printing is a process of making 3-dimensional solid objects from a digital file. Unlike creating an object from a plastic injection mold or pouring molten iron into a cast, 3D printing takes place by "printing" multiple and numerous 2D layers until the object is formed. One layer is added to the next layer over and over again, forming the object.[25]

The printing of an object starts before the 3D printer enters the process. The design begins as an idea that is translated into a design drawing, a blue print, through the use of a computer program that is referred to as a CAD drawing (Computer Aided Design). At the onset of this book, the authors' research was initially entered via a word processer before being printed. Likewise the 3D object is stored as a CAD model before getting sent to a 3D printer.[25] The three dimensional model is finally sent to the 3D printer along with the material that the object is to be made of: plastic, ceramic, or even human tissue. The layering is iterative until the actual object is created.[25]

3D printing is no longer a theoretical technology producing objects in a laboratory. 3D technology is being utilized in a wide variety of industries from the production of spare parts to the actual creation of weaponry like guns. A recent article in the tech section of the Huffington Post reported that the first ceramic, 3D printed gun was fired, and along with it a spark of debate on both sides of the issue.[13] High capacity ammo clips are now becoming commonplace.[10]

Figure 41: Ammo clip[10]

3D printers have increasingly become less costly to produce. Experts predict 3D printers will be common in homes in the coming years. 3D printing uses range from practical objects for everyday use, to commercial products and parts used in manufacturing.

Printed Organs

The 3D technology holds promise for bioprinting of human parts for medical purposes.[23] David Weldon in a report to FierceCIO quoted Debbie Holton, director of North American events and industry strategy for the SME (formerly the Society of Manufacturing Engineers) that "3D printing will inspire the next stem cell-style medical debate." "The day when 3D bio printings of human organs are readily available is drawing closer, and will result in a complex debate involving a great many political, moral and financial interests." Artificial or replacement tissue is commonly grown on collagen scaffolds [15] that contain biological starter cells. The end goal here is the growing of a biocompatible piece of tissue to repair or replace a patient's own damaged body part, such as bone, cartilage, blood vessels, or skin. [20, 23, 21]

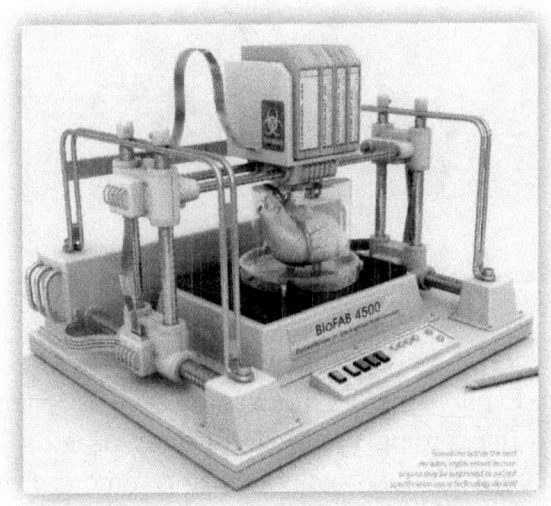

Figure 42. Bioprinting Equipment

Neuroscientists

Human sensory and motor systems provide the natural means for the exchange of information between individuals, and, hence, the basis for human civilization. An international team of neuroscientists and robotics engineers have demonstrated the viability of direct brain-to-brain communication in humans.[8,16] What is the potential for information that is available in the brain to be transferred directly in the form of the neural code, bypassing language altogether? In the journal, PLoS ONE, research has concluded that human brains can communicate directly with each other. In both cases, the transmitted pseudo-random sequences carried encrypted messages encoding a word – "hola" ("hello" in Catalan or Spanish) in the first transmission, "ciao" ("hello" or "goodbye" in Italian) in the second. [8,9,18]

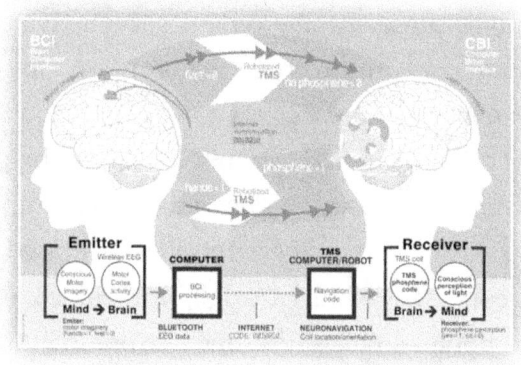

Figure 43. Brain-to-brain (B2B) Communication System Overview[8]

Consider the Hogan conjoined twins from British Columbia who share a neural bridge – the twins are able to communicate thoughts without speaking or even looking at each other. Though this anomaly occurred naturally as they developed embryonically, their ability demonstrates that physically this is possible between two individuals.

Neuroprosthetics

Unlike pacemakers, dental crowns or implantable insulin pumps, neuroprosthetics – devices that restore or supplement the mind's capacities with electronics inserted directly into the nervous system - change how we perceive the world and move through it. There are three main types of neuroprosthetics: sensory prosthetics; motor prosthetics; and cognitive prosthetics. Neuroprosthetics are implantable and designed to replace or improve a function that already is present in the central nervous system.[24] The most famous neuroprosthetic device is the cochlear implant, which bypasses the eardrum and directly stimulates the human auditory nerve, giving the power of hearing to those who lack it. The first cochlear implant was built in 1957, and today, these implants are used by over 100,000 people.[29] Rush Limbaugh, the controversial yet extremely popular talk show host, is one of the 100,000 recipients and successful users of this technological advancement.

Dr. Marcus, a professor of psychology at New York University, and Dr. Koch, the chief scientific officer at the Allen Institute for Brain Science, recently inquired:

> *What would you give for a retinal chip that let you see in the dark or for a next-generation cochlear implant that let you hear any conversation in a noisy restaurant, no matter how loud? Or for a memory chip, wired directly into your brain's hippocampus, that gave you perfect recall of everything you read? Or for an implanted interface with the Internet that automatically translated a clearly articulated silent thought ("the French sun*

king") into an online search that digested the relevant Wikipedia page and projected a summary directly into your brain?[28]

Well, many of these medical devices are either readily available, or will be in the not too distant future...

In the area of motor neuroprosthetics, there is the well-known pacemaker, which stimulates the heart to beat when the natural cluster of pacemaker cells is experiencing problems. Bladder control neuroprosthetics have assisted patients with paraplegia due to spinal cord damage. Motor neuroprosthetics for the conscious control of movement has received notoriety as well in recent years. Patients who are totally paralyzed can be given these implants, which allow them to control a computer mouse for spelling out messages, playing games, or surfing the web. For someone who is otherwise locked in an unresponsive body, these implants can radically improve quality of life.

For the first time ever, a paralyzed man can move his fingers and hand with his own thoughts thanks to an innovative partnership between The Ohio State University Wexner Medical Center and Battelle. Ian Burkhart, a 23-year-old quadriplegic from Dublin, Ohio, is the first patient to use Neurobridge™, an electronic neural bypass for spinal cord injuries that reconnects the brain directly to muscles, allowing voluntary and functional control of a paralyzed limb. Burkhart is the first of a potential five participants in a clinical study. [17]

Retinal prosthesis has progressed from the laboratory to the clinic over the past two decades. Currently, two devices have regulatory approval for the treatment of retinitis pigmentosa. These devices provide partial sight restoration and patients utilize this technology for improved vision in their everyday lives. Improved mobility and object detection are some of the more notable findings from the clinical trials. However, significant vision restoration will require advanced technology as well as an improved understanding of the interaction between electrical stimulation and the retina.[22] Research in visual neuroprosthetics has given rise to extremely fine electrodes (somewhat microscopic), thinner than a human hair. This has helped progress tangential areas of neurophysiology, but unfortunately true visual prostheses - devices that would allow the blind to see - are still in development. Scientists have observed that selective stimulation of the visual cortex allows subjects to see phosphenes - the little glowing blurs you see when you rub your eyes - in pre-determined areas of the visual field. Research has produced visual prostheses that give patients fuzzy vision with a pixel resolution of about 20 x 20, but these are just experimental and not ready for mass use. [27] Duke University reported that Larry Hester is the seventh person in the U.S. to have a so-called bionic eye - an Argus II Retinal Prosthesis Device. The report indicated that Dr. Hahn implanted the device that let Hester see for the first time in 33 years. Using wireless technology, a sensor is implanted in the eye to pick up light signals sent from a camera mounted on special eyeglasses.[6] The implant will not provide normal sight, but the patient will see light and shadows that previously were not possible.[3]

Neuroprostheses electrically stimulate paralyzed muscles to provide functional enhancement for individuals with neurological disorders, especially among persons with spinal cord injuries. Fully implanted neuroprostheses are reliable, require minimal maintenance and are user-friendly. These systems provide a variety of functions, including reaching, hand grasp and release, standing and stepping, bladder and bowel function and respiratory assist.

Chapter Wrap-up

Current technology - bionics, robotics, biology, and neuro sciences which exist today - provide the core for the potential existence of the DIGIPERSON. With the myriad of advancements to date, a major barrier exists in the implementation of implants in general – they are intrusive. To implant or connect to the brain, the only way of accomplishing this is by drilling into the skull. This is always accompanied by risk. Recent fabrication, actuation, and steering demonstrations of nanoscale robots represent the first crucial steps in overcoming the barrier. These medical nanobots can be used for targeted drug delivery and other uses like connecting circuitry in the brain itself.[30] Research has additionally supported the thesis that man's brain has adapted to the digital socio-technologic environment in which we exist. The plasticity of the brain has rerouted its neuro-circuitry. The brains of the younger generation are digitally hardwired often at the expense of the neuro-circuitry that controls one-on-one skills.[27] The DIGIPERSON has been evolved to a state that implants, transplants, biological changes, and neuro adaptations have already taken place. Further developments and innovations in the sciences are progressing at an alarming rate. Man's future is truly on a path to being a DIGIPERSON.

End Notes

1. Bhadra, N., & Chae, J. (2009). Implantable neuroprosthetic technology. *Neurorehabilitation*, 25(1), 69-83. doi:10.3233/NRE-2009-0500

2. BioFab 4500 retrieved March 11, 2015 from http://besuccess.com/2013/11/3dbioprinting/

3. Cantor, M. (2014). *After decades of blindness, man now sees with bionic eye.* Retrieved March 15, 2015 from http://www.foxnews.com/health/2014/10/15/after-decades-blindness-man-now-sees-with-bionic-eye/

4. Da Vinci (2013). *The da Vinci® Surgery experience.* Retrieved January 1, 2015 from http://www.davincisurgery.com/assets/docs/da-vinci-surgery-fact-sheet-en-1005195.pdf?location=1&version=b

5. Draxler, B. (2014). The KMatrix: *The bionic revolution begins. Popular Science,* December 2014.

6. Duke (2014). *NC's first bionic eye recipient sees for first time in 33 years.* Retrieved March 15, 2015 from http://www.dukemedicine.org/blog/#!/ncs-first-bionic-eye-recipient-sees-first-time-33-years

7. Graham, (2005). *Doctors perform surgery over the web.* Retrieved January 2, 2015 from http://pcin.net/update/2005/08/09/doctors_perform_surgery_over_the_web/

8. Grau, C., Ginhoux, R., Riera, A., Nguyen, T.L., Chauvat, H., et al. (2014). Conscious brain-to-brain communication in humans using non-invasive technologies. *PLoS ONE* 9(8): e105225. doi:10.1371/journal.pone.0105225

9. Grau, C., Ginhoux,R., Riera, A., Lam Nguyen, T., Chauvat, H., Berg, M., Amengual, J. L., Pascual-

301

Leone A., & Ruffini, G. (2014). Conscious brain-to-brain communication in humans using non-invasive technologies. *PloS ONE August 2014, Volume 9(8)*. Retrieved March 9, 2015 from http://www.plosone.org/article/fetchObject.action?uri=info:doi/10.1371/journal.pone.0105225&representation=PDF

10. Greenburg, A. (2014). *Gunsmiths 3D-print high capacity ammo clips to thwart proposed gun laws.* Retrieved January 1, 2015 from http://www.forbes.com/sites/andygreenberg/2013/01/14/gunsmiths-3d-print-high-capacity-ammo-clips-to-thwart-proposed-gun-law

11. IEEE Transactions on Biomedical Engineering - publication information. *Biomedical Engineering, IEEE Transactions (61)*3, pp.C2,C2, March 2014 doi: 10.1109/TBME.2014.2306304

12. Kandel, E. (2013). The new science of mind and the future of knowledge. *Neuron (80)*, 546–560.

13. Klienman, A. (2013). *The first 3D Gun has fired.* Retrieved January 1, 2015 from http://www.huffingtonpost.com/2013/05/06/3d-printed-gun-fired_n_3222669.html

14. Shi, L.Z., Berns, M.W., & Botvinick, E. (2008). "RoboLase": Internet-accessible robotic laser scissors and laser tweezers microscope systems. *Medical Robotics,* Vanja Bozovic (Ed.), ISBN: 978-3-902613-18-9, InTech, Available from: http://www.intechopen.com/books/medical_robotics/_robolase___Internetaccessible_robotic_laser_scissors_and_laser_tweezers_microscope_systems

15. Mertsching, H., Walles, T., Hofmann, M., Schanz, J., & Knapp, W. H. (2005). Engineering of a vascularized scaffold for artificial tissue and organ generation. *Biomaterials,26* (33), 6610-6617. doi:10.1016/j.biomaterials.2005.04.048.<http://www.sciencedirect.com/science/article/pii/S014296120500 3169>

16. Nicolelis, M.A. (2010) *Beyond boundaries: The new neuroscience of connecting brains with machines and how it will change our lives.* St. Martin's Griffin.

17. Ohio State University (2014). *New device allows brain to bypass spinal cord, move paralyzed limbs.* Retrieved November 1, 2014 from http://www.medicalcenter.osu.edu/mediaroom/rele ases/Pages/New-Device-Allows-Brain-To-Bypass-Spinal-Cord,-Move-Paralyzed-Limbs.aspx

18. Rao, R.P.N., Stocco, A., Bryan, M., Sarma, D., Youngquist, T.M., et al. (2014). A direct brain-to-brain interface in humans. *PLoS ONE 9*(11): e111332. doi:10.1371/journal.pone.0111332

19. "RoboLase": Internet-accessible robotic laser scissors and laser tweezers microscope systems. Beckman Laser Institute, University of California, Irvine, Deparment of Biomedical Engineering, University of California, IrvineUSA1 IEEE TRANSACTIONS ON BIOMEDICAL ENGINEERING, VOL. 60, NO. 3, MARCH 2013 693

20. Smith, L. A., & Ma, P. X. (2004). Nano-fibrous scaffolds for tissue engineering. *Colloids And Surfaces B (Biointerfaces), 39*(3), 125-131. doi:10.1016/j.colsurfb.2003.12.004.< http://www.sciencedirect.com.uri.idm.oclc.org/scie nce/article/pii/S0927776503003035

21. Thilmany, J. J. (2012). Printed life. *Mechanical Engineering,* *134*(1), 44-47. [image] < http://largecontent.ebsco-content.com.uri.idm.oclc.org/embimages/2f1c0c45d cdf63797ceb4dbee67d95bc/54397580/rdk/mee/01jan 12/45n1.jpg>

22. Weiland, James D., & Humayun, M.S. *IEEE Transactions on Biomedical Engineering.* May2014, Vol. 61 Issue 5, p1412-1424. 13p. DOI: 10.1109/TBME.2014.2314733.

23. Weldon, D. (2014). *3D printing could revolutionize medical prosthetics, SME says.* Retrieved March 11, 2015 from http://www.fiercecio.com/story/3d-printing-could-revolutionize-medical-prosthetics-sme-says/2014-02-04

24. *What are neuroprosthetics?* Retrieved March 11, 2015 from http://www.wisegeek.com/what-are-neuroprosthetics.htm

25. *What is 3-D printing?* Retrieved March 11, 2015 from http://3dprinting.com/what-is-3d-printing/

26. Carr, N. (2008). *Is Google making us stupit.* TheAtlantic.com retrieved March 10, 2015 from http://www.theatlantic.com/magazine/archive/2008 /07/is-google-making-us-stupid/306868/

27. Small, G., & Vorgan, G. (2008). *iBrain: Surviving the technological alteration of the modern mind.* New York: Harper Collins.

28. Marcus, G., and Koch, C. (2014). The future of brain implants. *The Wall Street Journal.* Retrieved March 1, 2015 from

http://www.wsj.com/articles/SB10001424052702304 91490457943559298178 0528

29. Kumari, R., & Kumari, A. (2009). *Neuroprosthetics.* Retrieved March 14, 2015 from http://www.asctbhopal.com/visiontechsouvenir200 9/cyberworld/neuroprosthetics.pdf

30. Kroeker, K. (2009). Medical nanobots. *Communications of the ACM, 52(9).*

31. Dominus, S. (May 25, 2011). *Can conjoined twins share a mind?* Retrieved June 18, 2015 from http://www.nytimes.com/2011/05/29/magazine/cou ld-conjoined-twins-share-a-mind.html?_r=0

Epilogue

An odyssey can be defined as a long series of wanderings or adventures, especially when filled with notable experiences, hardships, etc.[1] Consider all of the different sociological and technological meanderings encountered during this odyssey detailed herein. Along the way, the personal interactions have transformed significantly from a mere 10-20 years ago, while the attendant social and business constructs have transformed before our very eyes!

In 2001, Jim Collins authored a book called *From Good to Great* where he chronicled the development of relatively *"good"* companies into *"great"* companies.[2] A key conclusion that was drawn from that substantive research was that organizations which painstakingly developed the right people, processes, and strategies were able to make it look easy (after the initial hard work). This was termed the *"flywheel"* effect because the research concluded that once the hard work was accomplished, it became easier for the organization to continually achieve sustained success-this premise was analogous to a *flywheel* that requires a great deal of inertia to get it started, but then becomes increasingly easier due to the attendant momentum. With this in mind, consider all of the inertia required for our current socio-technological odyssey chronicled herein...is the *flywheel* gaining momentum?

One has to wonder what the potential impacts are going to become for our DIGIPERSON? Obvious positive impacts are occurring through health breakthroughs, ease of communication, ease of data transfer, etc. But, what consequential impacts lie in store for these supposed benefits as related to our interpersonal relationships, communication and ongoing wellbeing? This is not intended as an anti-treatise for the rapidly evolving technological infrastructure and attendant applications; rather, it is a harbinger for the leverage of all the positive benefits while being an omen to proactively *"manage"* this transformational evolution (or is it a revolution?).

Centuries down the road our physical presence will be unearthed in all of the data nuggets we have dropped along the way; the curiosity for future archaeologists will lie in the data itself, and not necessarily the brick and mortar we have left behind. Wouldn't it be fun to see these adventurers in action trying to make sense of the plethora of data they have "unearthed"? So, the *flywheel* is spinning; our socio-technological odyssey continues, and the maturation of the DIGIPERSON moves on towards speciation. Next stop?

We hope you have enjoyed this snapshot of the socio-technological odyssey and thank you for joining us on the journey. Watch your step as you exit this ride!

The odyssey continues and will include many new features, some alarming. A sneak peek includes:

- Cyber warfare
- Drone armies
- Holographic TV
- Truly self-driving transportation
- Smart cities and roads
- Implanted personal identity chips
- Ever present surveillance
- Nano medicine
- Learning machines/Artificial intelligence
- Exploring galaxies "far far away"

The next ride is leaving soon.
Dick & Jerry

Final Endnotes

1. http://dictionary.reference.com/browse/odyssey?s=t
2. Collins, J. (2001). *Good to great: Why some companies make the leap--and others don't.* New York: HarperBusiness.

Appendix A: Previously Published Research

Ponschock, R.L,. & Becker, G.F., (2014). Threat Vectors on Privacy: The Trail of "Digital Identity Crumbs©" *Journal of Strategic and International Studies Volume X Number 1 2015 ISSN 2326-3636*
Presented at 2015 Multidisciplinary Academic Conference (InstSIS World) Miami, Fl.

Ponschock, R.L,. & Becker, G.F., (2014). Digital Reputations: A Paradigm Shift for Individuality and Business. *Journal of Academy for Advancement of Business Research. ISSN: 233222-0311.*

Ponschock, R.L,. & Becker, G.F., (2013). Digitization of Society: The Avatar Effect. *Review of Strategic and International Studies. ISSN: 2326-8085.*

Ponschock, R.L,. & Becker, G.F., (2012). Digitization of Society: The Continuum to a Speciation Event. *Review of Business Research. ISSN: 1546-2609.*

Ponschock, R.L,. & Becker, G.F., (2011). CLOUD Technology: A Transformational Dynasty on the ICT evolutionary continuum and contemporaneously a societal speciation event. *European Journal of Management. ISSN: 1555-4015.*

Ponschock, R.L,. & Becker, G.F., (2010). Virtuality of boundaries: - the iceberg in the current business curriculum: A mandate for systemic thinking. *Association for Global Business Proceedings, 22, Paper 29. ISSN: 1050-6292.*

Becker, G.,F., & Ponschock, R.L., (2009). Blurred boundaries: A case for systemic thinking in the business curriculum. *International Journal of Business Research, 9(,1), ISSN 1555-1296.*

Ponschock, R., Greif, T.B. and St John, L.(2008). Transformative change through virtualization in a globally networked economy. *Proceedings of International Academy of Business and Economics (IABE)*

Harris, M. E., Becker, G. F., Hallcom, A. S., & Ponschock, R. L. (2008). Success as the driver in the socio-economic equation trilogy: Part 1. *European Journal of Management, 8(3),* 55-60. (ISSN: 1555-4015).

Harris, M. E., Becker, G. F., Hallcom, A. S., & Ponschock, R. L. (2008). Energizing the leadership in the socio-economic equation trilogy: Part 2. *European Journal of Management, 8(4),* 76-80.(ISSN: 1555-4015).

Harris, M. E., Becker, G. F., Hallcom, A. S., & Ponschock, R. L. (2008). Managing the complexity & chaos in the socio-economic equation trilogy: Part 3. *European Journal of Management, 8(4),* 71-75. (ISSN: 1555-4015).

Harris, M. E., Hallcom, A. S., Becker, G. F., Greif, T. B., & Ponschock, R. (2008). Why is leveraging transformative change critical in management consulting today? *Paper presented at Academy of Management Annual Conference in Anaheim, CA.*

Harris, M. E., Becker, G. F., Hallcom, A., & Ponschock, R. L. (2008). *Mastering a New Paradigm in Management Consulting (2nd ed.)*. Createspace. (ISBN: 978-1438250496)

Ponschock, R., & Greif, T.B. (2008). Digital age archeology: The social impact of technology imprinting. *Proceedings of the 2nd International Multi-Conference on Society, Cybernetics and Informatics, 3, 120-126*. (ISBN-13: 978-1-934272-48-0).

Ponschock, R., & Greif, T.B. (2007). Archaeological Excavating in Virtual Villages: A primer on discovery of artifacts from a digital culture. *Proceedings of International Academy of Business and Economics (IABE)*

Figures

Images

Image 1: www.computerhistory.org

Image 2: www.thecorememory.com/html/ncr_8250.html

Image 3: cute-baby-images.clipartonline.net/_/rsrc/1362073304916/stork-carrying-baby boy/Stork-Carrying-Baby-Boy_8.png?height=320&width=320&height=400&width=400

Image 4: Drone image: Amazon.com

Image 5: www.economist.com

Image 6: www.dreamstime.com E-waste

Image 7: www.pintrest.com *Six Million Dollar Man*

www.ingramcontent.com/pod-product-compliance
Lightning Source LLC
Chambersburg PA
CBHW051442170526
45166CB00001B/82

* 9 7 8 1 5 1 9 4 9 3 5 0 7 *